RAND PROJECT AIR FORCE

Cyber Practices

What Can the U.S. Air Force Learn from the Commercial Sector?

Lara Schmidt, Caolionn O'Connell, Hirokazu Miyake, Akhil R. Shah,
Joshua William Baron, Geof Nieboer, Rose Jourdan, David Senty,
Zev Winkelman, Louise Taggart, Susanne Sondergaard, Neil Robinson

Prepared for the United States Air Force
Approved for public release; distribution unlimited

For more information on this publication, visit www.rand.org/t/rr847

Library of Congress Cataloging-in-Publication Data
ISBN: 978-0-8330-9032-4

Published by the RAND Corporation, Santa Monica, Calif.

Preface

This document describes commercial practices for cyber workforce management and organizational issues as determined by interviews with carefully selected organizations and a broad survey of the literature. The commercial practices described here are intended to inform the U.S. Air Force (USAF) as it endeavors to continually improve the management of its cyber forces. We describe the basis for the existence of these practices, the benefits they convey, and their applicability to USAF.

The research reported here was conducted within a fiscal year 2014 project entitled Best Practices to Inform USAF Cyber Squadrons of the Future, sponsored by Maj Gen Earl Matthews, director of Cyberspace Operations at Headquarters USAF, and conducted within the Force Modernization and Employment Program of RAND Project AIR FORCE.

RAND Project AIR FORCE

RAND Project AIR FORCE (PAF), a division of the RAND Corporation, is the U.S. Air Force's federally funded research and development center for studies and analyses. PAF provides the Air Force with independent analyses of policy alternatives affecting the development, employment, combat readiness, and support of current and future air, space, and cyber forces. Research is conducted in four programs: Force Modernization and Employment; Manpower, Personnel, and Training; Resource Management; and Strategy and Doctrine. The research reported here was prepared under contract FA7014-06-C-0001.

Additional information about PAF is available on our website:
http://www.rand.org/paf/

Contents

Figures

Tables

Summary

To meet the challenges of the cyberspace era—including the rapid rate of change in technology, the growing cyber threat, and the need to integrate cyber operations with operations in other warfighting domains—the U.S. Air Force (USAF) must find effective ways to organize, train, and equip its cyber forces. Progress on these issues has been made over the past decade and continues with the maturation of United States Cyber Command. However, the criticality of cyber missions has led USAF to seek further improvements. As such, USAF asked RAND Project AIR FORCE (PAF) to assist in identifying improved approaches to cyber organizational and workforce issues. Specifically, this report describes our efforts to identify successful processes and practices from the commercial sector that might be applicable to USAF.

To identify successful commercial practices, we took a twofold approach—a wide-ranging literature review and interviews with a carefully crafted set of commercial organizations, selected for their similarities to USAF and for their reputations of cyber excellence. Companies were identified to be similar to USAF in size, cyber functions performed, exposure to cyber threats, and operational environment. We found strong parallels in the commercial sector for Department of Defense information network operations (DoDIN Ops) and defensive cyber operations (DCO). Although none of the companies we interviewed were as large as USAF or required to function in deployed and contested operating environments, the commercial practices we describe might provide effectiveness and efficiency gains and, at the very least, are informative for USAF.

Several Commercial Cyber Practices Are Informative for USAF

In this report, we identify commercial practices for cyber workforce management that are in widespread use and provide effectiveness and efficiency gains to the organizations that employ them. USAF should find these practices informative because of its similarities to the organizations we interviewed. Furthermore, we found corroboration for the benefits of these practices from the academic literature. Next, we summarize four top-level practices and describe how USAF might choose to apply them to cyber workforce management.

Practice #1: Information Technology and Information Security Are Managed as Two Separate Disciplines

Information technology (IT) consists of tasks related to providing cyber services, including the operation and maintenance of computer systems, networks, and data; requirements planning; knowledge management and in-house software and systems development; computer user and network support; and software assurance. Information security (InfoSec) consists of tasks related to protecting and defending systems and networks; detecting, investigating, and responding to security incidents; ensuring information assurance compliance; performing security systems development; and designing system security architectures.

We find that practices for the organization, training, and development of IT staff throughout their careers differ from practices for the organization, training, and development of InfoSec staff. However, the organizations we interviewed repeatedly emphasized that one field is not better than the other, nor is one a natural progression from the other. While these fields share some common baseline of knowledge—similar to chemists and chemical engineers trained in some of the same fundamentals—the skill sets, work styles, job roles, and management strategies that apply to the two fields are different. As such, commercial practice is to manage these two fields differently.

IT practices include the use of a hierarchical organization. The head of the IT organization, the chief information officer (CIO), empowers only a few senior managers to make decisions with regards to the company's IT priorities and enforces adherence standardized processes, when applicable, for such tasks as configuring hardware or help desk support. Personnel are organized into functionally aligned groups (e.g., network architects, customer support) with relatively more staff per supervisor when compared with InfoSec. Training focuses on topics related to development, operations, and the maintenance of systems, as do job roles. Staff typically progress through an entire career in the IT discipline area, either as technical experts or eventually transitioning into IT management. Rarely, a staff member might transition to InfoSec, but this is the exception, not the rule.

In contrast, InfoSec practices include more-decentralized organizations. The head of the InfoSec organization, the chief information security officer (CISO), empowers more people throughout his or her organization to make security decisions that allow for rapid responses and decisionmaking, which is often required in InfoSec. Personnel are organized in small, cross-functional teams aligned by mission (e.g., security incident response, risk assessment). InfoSec organizations assign fewer staff per supervisor to compensate for the complexity and dynamics of the operating environment. Furthermore, some InfoSec jobs can require longevity of five or more years before personnel truly master the duties. Therefore, InfoSec career progression can be more deliberate, and it is rare for an InfoSec staff member to transition to IT.

We found that the commercial practice of managing the IT and InfoSec disciplines as two separate fields is likely to be informative for USAF. There are widespread similarities between USAF and the commercial sector regarding the IT and InfoSec functions performed and the operational conditions experienced. Were USAF to adopt the practice of managing IT and InfoSec as distinct disciplines, we expect that the benefits would be reflected in the increased technical depth of personnel, who would yield increased effectiveness and efficiency. Possible reductions in the total size of the cyber workforce might be expected, although not to the extent found in the commercial sector, since USAF must retain some redundant capabilities to ensure resilience in the face of wartime threats and to adequately address the needs of the regional combatant commands (RCCs) and their air components.

Practice #2: IT Is a Critical Core Function Performed by a Large In-House Staff

Although the size of an IT department varies by industry sector, we observed approximately one IT staff member per 25 employees for companies most similar to USAF. Furthermore, we found striking similarities across the companies we interviewed regarding the levels of effort allocated to InfoSec relative to IT, and these values were supported in the literature. Companies maintained approximately 20 times more IT personnel than InfoSec personnel—i.e., approximately 95 percent of a company's cyber personnel are devoted to IT, and only 5 percent are aligned with InfoSec. Additionally, this IT workforce consists largely of in-house personnel.

IT outsourcing was commonly pursued for tier 1 help desk or desktop services because these functions were not considered part of most companies' core competencies. However, other IT functions (e.g., data administration, knowledge management, network services, system administration, systems security analysis, system design, requirements analysis, user account management) were considered critical core capabilities that companies wanted to manage internally. This sentiment is echoed in independent surveys of senior executives who reported that they could endure less than a day of downtime from their IT systems before the disruption became serious enough to jeopardize the survival of their entire company. The criticality of IT drives corporate practice to maintain robust, productive, in-house IT workforces. This practice should be informative to USAF as it considers the benefits and risks of outsourcing.

Furthermore, IT personnel are consolidated under a single corporate-level organization, headed by the CIO, as opposed to reporting to heads of business units.[1] This consolidation delivers efficiencies by eliminating the duplication of effort and reinforcing the skills of individuals through their close collaboration with peers. While USAF does not currently employ such consolidation of IT, even if it did, we would not expect USAF to reap the level of efficiency gains observed in the commercial sector (i.e., ratios of one IT staff member per 25 employees). This is because USAF must retain some redundancies to ensure that it can deliver services in cyber-contested warfighting environments at operating locations around the world. However, USAF might be able to find efficiencies above current levels by applying consolidation to the extent possible given warfighting constraints. In particular, consolidation mirroring that found in large multinational conglomerates is a model applicable to USAF. These conglomerates, with multiple subsidiaries in different countries and/or industry sectors, consolidate only *within* country or sector to ensure maximal efficiency gains while maintaining networks in different

---

[1] Note that consolidation does *not* imply geographic colocation. In fact, corporate practice is that IT staff are present at all the major operating locations to deliver tailored support. However, they report to the CIO, not the head of the operating location. In many instances, however, there are liaisons between the CIO and heads of other business units to coordinate IT needs and service provision.

regulatory spheres. For example, a conglomerate might maintain a CIO atop a consolidated IT organization at the UK subsidiary instead of consolidating further across the entire multinational conglomerate due to nation-specific laws. USAF might be able to apply this model to consolidate IT within, but not across, operational boundaries (e.g., RCCs or major commands [MAJCOMs]) to ensure that warfighting needs could be met while still gaining economies of scale within a region or functional regime.

Practice #3: Technical Leadership Is Valued and Cultivated

Commercial practice suggests that IT and InfoSec managers need to be well versed in their field to manage effectively, accurately judge the quality of their staff's work, and make informed decisions. This leadership has the additional challenge of keeping up with the rapid pace of technological change. Technology skills—such as programming and scripting, as well as knowledge of hardware—are highly perishable. Commercial practice tackles this challenge by allowing staff and leaders to specialize within either IT or InfoSec, thereby reducing the universe of possible trends with which they must keep pace. Second, managers are required to stay current through recurring training and regular hands-on opportunities in their departments. While managers were not as efficient as their staffs at performing such hands-on tasks, managers reported that they periodically performed these tasks specifically to remain current and, therefore, perform better as managers.

Commercial practices related to organizational strategies, training opportunities, and the management of career progression all support the development of leaders with technical depth. Consolidated organizations encourage the cross-flow of knowledge among staff, building technical depth in ways not possible in smaller, more isolated organizations. Perhaps more important, however, is the practice of progressing staff and leadership in careers within a single discipline—either IT or InfoSec. Allowing this specialization enables leaders to master disciplines at a manageable speed before broadening to other areas of that discipline.

USAF could apply practices associated with organization and recurring training to increase the technical depth of its cyber leaders. As described in the previous section, consolidating IT staff to the extent feasible under a (perhaps regional or MAJCOM aligned) CIO and establishing strong linkages among staff of like function across geographical locations should improve technical depth, as would increasing the frequency of training. However, a fundamental question is *whether* USAF should value technical depth of leadership in the first place. One argument against doing so is the need for cyber leaders to remain competitive for promotion, which favors the creation of generalists capable of leading any cyber organization while on a path toward senior leadership. However, there might be ample opportunity for developing the breadth required for very senior positions using a more gradual approach that still reinforces technical depth. The commercial practice for retaining technical depth while developing breadth is to provide experiences that allow individuals to apply their specialty in other parts of the company. For example, a software security engineer based in the corporate headquarters of an InfoSec organization might be broadened by an assignment to lead the security aspects of a high-profile development program in one of the business units. This approach serves several purposes—it retains depth (in this case, in InfoSec) by applying the employee's specialty to a new application, it exposes the employee to the needs of the business units, it demonstrates the employee's value to the company, and it aids retention by keeping top-performing employees challenged, working on interesting projects, and progressing in their careers. This practice, which encourages technical depth in leadership, might be informative to USAF promotion and career field management practices.

Practice #4: Traditional Approaches Are Employed for Recruiting and Retention

Like USAF, many of the large companies we interviewed preferred to hire early-career staff and retain them for decades. Commercial practice is to hire staff with a bachelor's degree, which provides a strong foundation of relevant knowledge and demonstrates an ability to succeed in a professional setting. Companies usually recruit these graduates from reputable colleges with science, technology, engineering, and mathematics (STEM) degrees, especially computer science, InfoSec, IT, computer engineering, and electrical engineering.

Unlike the commercial sector, the vast majority of the enlisted force are not required to have college degrees.[2] Therefore, USAF favors substantially more-rigorous selection criteria than the commercial sector to vet nondegreed applicants. However, some of the more cutting-edge cybersecurity companies have instituted cyber aptitude or skills testing as part of the application process to evaluate a candidate's expertise or mind-set for a particular discipline. Companies also looked for a personal interest in and affinity for cyber, such as participation in cyber competitions during secondary school and contribution to open source or ethical hacker forums. In fact, commercial practice for elite InfoSec jobs employed such tests of aptitude and ability *in addition to* formal educational requirements. To provide additional assurance that enlisted accessions will become productive members of the cyber ranks, USAF is in the process of adopting similar types of aptitude tests.

Finally, corporations actively managed their cyber workforce and instituted policies to maximize retention of their skilled personnel. Corporations reported that those retention policies were successful, with most companies largely satisfied with their low attrition rates. However, exorbitant salaries are not viewed as the secret to retention. In fact, median salaries for corporate IT and InfoSec professionals are similar to the pay and benefits for military personnel, when accounting for additional allowances and tax advantages. Instead, corporate practice focuses on providing job satisfaction through good working environments, belief in the mission, opportunities for training, exposure to and engagement with professional organizations, and

---

[2] USAF is moving to a STEM degree requirement for cyber officers.

access to interesting assignments. These commercial practices are directly applicable to USAF and, in many cases, are similar to how USAF retains good personnel today.

One exception to the above practices relates to the most elite InfoSec professionals, those with unique skills that few possess. These "ninjas" are the competitive advantage for cutting-edge cybersecurity firms and are increasingly in demand in other corporate settings. The relative scarcity of these skill sets allows qualified individuals to command high salaries. USAF might similarly find personnel with these unique skills to be worthy of retention programs not offered to the majority of the cyber workforce.

Options for USAF to Implement Applicable Commercial Practices

Although the purpose of this research was not to analyze how USAF should implement commercial practices, we identified ways in which commercial practices might be applicable to USAF and, therefore, merit further investigation. Next, we describe several promising options for tailoring commercial practices to USAF that might help improve USAF cyber workforce practices. The applicability and rationale for each of these recommendations is described in detail in the report. We recognize that there might be other considerations that would preclude USAF from adopting these practices; however, we offer them as specific issues to investigate.

Outsourcing

Commercial companies limit the outsourcing of many cyber functions they regard as core capabilities, critical for the functioning of their business. When companies lack the capability to perform a core cyber function, they might design an outsourcing arrangement that not only accomplishes the function but also increases the expertise of their in-house staff, to later insource this core function. USAF should identify which IT and InfoSec functions are critical to USAF missions and carefully assess the risk of outsourcing—whether to contractors or other government providers—those core capabilities.

Workforce Size

In the commercial sector, the IT workforce remains large, while InfoSec continues to grow slowly—current ratios show 95 percent of the cyber workforce engaged in IT and 5 percent in InfoSec. If, like commercial practice, USAF decides to limit outsourcing, USAF similarly will need to maintain a large IT force into the future. Current comparisons with commercial practice indicate that the USAF defensive cadre might need to be larger, particularly given that USAF's mission might result in additional demands on its DCO forces not present in commercial companies. Therefore, USAF might require an InfoSec force greater in size proportionally than seen in commercial entities. USAF should conduct a workforce study to determine the appropriate ratio of IT to InfoSec in the USAF's cyber workforce. In the meantime, the ratio derived from commercial practice could be considered a lower bound for USAF force structure planning.

Technical Depth

Commercial companies depend on personnel with deep technical expertise developed over long periods. Particularly for complex jobs (e.g., cyber emergency response team members), longevity is required to develop even the basic levels of expertise required to function effectively. Yet commercial practice values longevity even for less complex jobs, since efficiency gains can be realized by keeping staff deeply rooted in their specialty. Longevity is one mechanism the commercial sector uses to function despite constant pressures to be lean and control costs. USAF should evaluate the extent to which some IT and InfoSec roles could be filled by civilians, guard, and reserve personnel to increase longevity in these positions, thereby increasing technical depth. Associating highly technical InfoSec jobs in a way that either ensures longevity in these positions or rotates personnel among very similar positions, allowing for maximal transfer of knowledge, might deliver both efficiency and effectiveness gains for USAF.

Organizational Design

As opposed to many small IT organizations, each supporting a business unit, commercial practice has been to consolidate these staff into larger organizations, each supporting a larger portion of the company (e.g., one IT organization supporting an entire subsidiary), to reap efficiencies and other benefits. Formal liaisons are assigned between the consolidated IT organizations and the supported units to ensure that the specialized needs of user communities are met. Like the commercial sector, consolidating USAF IT operational organizations into larger organizations (potentially regional or MAJCOM aligned) might improve customer support while gaining efficiency. USAF might analyze such IT consolidation in conjunction with the implementation of the Air Force Installation and Mission Support Center.

Conversely, organizational design and commercial practice indicate that InfoSec groups excel as small, cross-functional teams associated with a mission. USAF appears to be applying this approach to InfoSec organizational design.

Career Field Management

Commercial practice consistently manages IT and InfoSec as distinct career fields. By doing so, companies report benefits to both effectiveness and efficiency. Furthermore, commercial practice does not support the notion that mastering IT is a prerequisite to performing well at InfoSec—staff typically enter into and progress through a career in one field. While a cursory evaluation indicates that many USAF enlisted career fields are aligned to one specialty, other enlisted and officer specialties appear to require an individual to master both fields. Given the rapid pace of technological change and the complexity of many of these activities, USAF should analyze the benefits of aligning Air Force Specialty Codes to *either* IT or InfoSec.

Accessions

Commercial companies indicated that passion and aptitude for IT and InfoSec produced the most-effective workers. Preferred candidates are those who demonstrate passion through their activities (e.g., ethical hacker certifications, participation in open source communities) and aptitude through their credentials (e.g., a bachelor's degree in information systems). Without a degree requirement for enlisted accessions, USAF should investigate the feasibility of establishing *tests* for IT aptitude and InfoSec aptitude as part of the enlisted accession process. Additionally, as part of the officer accession process, USAF should prefer candidates with relevant academic degrees from universities with noted cyber programs (e.g., National Security Agency and Department of Homeland Security cyber centers of excellence) or relevant extracurricular activities.

Reporting Relationships

The commercial sector employs a clear and effective command structure, wherein the CIO is responsible for IT operations and the CISO is wholly accountable for InfoSec operations. USAF's current command structure is far more complex. USAF should evaluate the role of the CISO within its organization and determine whether that person is sufficiently empowered to make security decisions that are weighted in the context of operational risk.

Acknowledgments

The research questions addressed in this report were conceived by Major General Earl Matthews, the director of Cyberspace Operations within the Office of Information Dominance and Chief Information Officer (HQ USAF A3C/A6C). We are indebted to his foresight and guidance throughout this project. We would also like to acknowledge the contributions of his team, Col (ret.) Jodine Tooke, Lt Colonel Dave Canady, and Chief Master Sergeant John Sanders, who provided context and guidance as the work progressed. Mr. Peter Kim, deputy director of Cyberspace Operations, was instrumental in USAF efforts to act on the findings of this work, and we thank him for his continued collegiality.

We are deeply indebted to the numerous representatives from the commercial corporations who agreed to take part in this effort. Their generosity of time and dedication to their professions and employers were truly inspiring. We thank them for reflecting on their trials and tribulations and for allowing us the benefit of those insights.

In particular, we would like to acknowledge the unique insight provided by former Air Force officers who are now employed in the private sector, such as Lt Gen (ret.) Charles Croom (vice president of cyber strategy and government relations, Lockheed Martin), Mr. Richard Bejtlich (chief security strategist, FireEye), Mr. Sebastian M. Convertino II (vice president of information security and strategic operations, CrowdStrike Inc.), and others who preferred anonymity. Additionally, we would like to express our gratitude to Mr. Brad Myers (Lockheed Martin) and Mr. Lee Holcolm (retired from Lockheed Martin).

We appreciate the opportunity to have helpful discussions with Air Force colleagues, including Ms. Kim Kendall (SAF/AQI), Lt Col Jack Jurgensen (AF/A6), Lt Col Travis Howell (JS/J3), Lt Col Basballe (AF/A3C/A6C), Lt Col Carlos Alford (SAF/CIO A6), Lt Col Christian Basballe (A3CXW/A6CXW), and Capt Jeff Crepeau (PACAF/A3CC). We benefited from the insights of Col Bradley Pyburn, commander, 624th Operations Center; Lt Col Glenn Garay, commander, 502nd Communications Squadron; and Maj Jason Parker, chief, Cyberspace Officer Assignments at Air Force Personnel Center; and Special Agent Justin Vallese of the FBI.

Our RAND colleagues from the Manpower, Personnel, and Training Program and the economics, statistics, and sociology departments served as great resources to us: Ray Conley, Lawrence Hanser, James Hosek, Lisa Harrington, Kirsten Keller, Maria Lytell, and Michael Kennedy. We would also like to thank our colleagues for collaborating on cyber topics, including: Ryan Henry, Isaac Porche, Michelle Ziegler, Lillian Ablon, Martin Libicki, and Don Snyder. In particular, we wish to acknowledge the suggestions provided by our reviewers, Daniel Ginsberg, Laura Werber, and John Davis, which made for a greatly improved final document.

Many thanks as well to Hon. Richard Danzig for encouraging this work and for all his efforts to bring sound analysis to cyberspace.

Finally, we are grateful for the able assistance of Karen Edwards and Patty Hazzard for helping this report come to fruition.

Abbreviations

ACM	Association for Computing Machinery
AFSC	Air Force Specialty Code
CEO	chief executive officer
CERT	Computer Emergency Response Team
CFO	chief financial officer
CIO	chief information officer
CISO	chief information security officer
CISSP	Certified Information Systems Security Professional
CRO	chief risk officer
CSIRT	Computer Security Incident Response Team
DCO	defensive cyber operations
DHS	Department of Homeland Security
DISA	Defense Information Systems Agency
DoD	Department of Defense
DoDIN Op	Department of Defense information network operation
IEEE	Institute of Electrical and Electronics Engineers
InfoSec	information security
IP	Internet protocol
$(ISC)^2$	International Information Systems Security Certification Consortium
IT	information technology
ITAA	Information Technology Association of America
JIE	Joint Information Environment
MAJCOM	major command
MCSE	Microsoft Certified Solutions Expert
MSSP	managed service security provider
NICE	National Initiative for Cybersecurity Education

NIST	National Institute of Standards and Technology
NSA	National Security Agency
OCO	offensive cyber operations
PCS	permanent change of station
PD	protect and defend
RCC	regional combatant command
SCADA	supervisory control and data acquisition
STEM	science, technology, engineering, and mathematics
USAF	U.S. Air Force

1. Introduction and Methodology

Cyber is pervasive in modern society, with individuals, corporations, and governments increasingly networked to the global public information infrastructure, taking advantage of advancements in digitization. These participants in cyberspace have seen increases in both efficiency and effectiveness as a result of their use of cyber systems, but they also find themselves confronting growing cyber threats that jeopardize the confidentiality, availability, or integrity of their data and, potentially, even their ability to do business. In fact, a recent survey characterizes cyber-related failures as an existential threat to companies that rely on cyber systems.[3]

The U.S. Air Force (USAF) relies on cyber systems for global air and space operations and for its business processes. USAF also conducts offensive and defensive cyberspace operations as part of joint military operations. To meet the challenges of the cyberspace era—including the blistering rate of change of technology, the growing cyber threat, and the need to integrate cyber operations with operations in other warfighting domains—USAF must find effective ways to organize, train, and equip its cyber forces. Progress on these issues has been made over the past decade and continues with the maturation of United States Cyber Command. However, the criticality of cyber missions has led USAF to seek further improvements.

USAF asked RAND Project AIR FORCE (PAF) to assist these efforts by identifying approaches to improve cyber organizational constructs and mitigate potential cyber workforce issues. Specifically, this report summarizes our efforts to identify successful practices from the commercial sector that might be applicable to USAF. In this report, we describe commercial practices associated with organizing; force sizing; staff recruitment, training, and retention; leadership; and outsourcing. However, two principal challenges present themselves when conducting such an analysis.

First, not all commercial practices are applicable, since USAF faces constraints not common in the commercial sector (e.g., a complex and extremely large global organization, a limited pool from which to recruit, an emphasis on effectiveness over efficiency). Second, successful commercial practices might not be unique or well founded (i.e., there might be many ways to achieve the same levels of efficiency and effectiveness; or a practice that experts deem to be "successful" or "best" might not necessarily be so). To ensure that the commercial practices we highlight in this report are worthy of consideration by USAF, we implemented the rigorous methodology described below.

---

[3] Economist Intelligence Unit, *Business Resilience: Ensuring Continuity in a Volatile Environment*, London: The Economist, 2007.

Methodology

We took a twofold approach to establish successful cyber practices from the commercial sector. We conducted a wide-ranging literature review to identify recommended cyber practices. Concurrently, we interviewed a set of commercial organizations, which were selected for their similarities to USAF and for their reputation of cyber competence (or, in many cases, cyber excellence). Only practices that were both pervasive in the companies we interviewed and had support from the literature were considered.

Literature Review

Our study reviewed academic and business literature, as well as military publications, on topics associated with cyber education, workforce, organization, and outsourcing. We also assessed industry themes, as described by surveys and analyses by government organizations, consulting firms, cybersecurity companies, and trade magazines directed toward chief information officers (CIOs) and chief information security officers (CISOs).[4] Given that cyber is a rapidly evolving field, we generally selected research conducted within the past ten years to ensure that we captured the most current practices.

A major component of our literature review informs the application of organizational design to cyber functions. *Organizational design* is the macro examination of organizations—social entities that are goal directed, deliberately structured and designed, and linked to the external environment.[5] That is, organization design helps determine the appropriate organizational structure to achieve a given objective.[6] A significant amount of academic research exists on this topic, but little of this work has been applied to cyber missions. This report applies those established techniques to cyber operations to determine the appropriate structure for cyber organizations.

Semistructured Interviews

We conducted a series of semistructured interviews with senior cyber personnel at companies and other government organizations that shared similar characteristics with USAF. We interviewed 26 companies and organizations from a wide range of industry sectors, including financial institutions, major manufacturing firms, defense industrial base firms, energy

---

[4] The CIO is the most senior executive in an enterprise responsible for the information technology and computer systems that support enterprise goals. The CISO is the senior-level executive within an organization responsible for ensuring that information assets and technologies are adequately protected.

[5] Richard L. Daft, *Organization Theory and Design*, 10th ed., Mason, Ohio: South-Western Cengage Learning, 2008, p. 11.

[6] Robert Duncan, "What Is the Right Organization Structure? Decision Tree Analysis Provides the Answer," *Organizational Dynamics* 79, no. 7 (1979).

companies, network security specialists, and large government agencies.[7] Of the 26 interviews, only 22 companies and organizations contributed to our analysis. Those organizations that were not included had limited applicability to USAF, as they predominately specialized in forensics. Of the 22 companies that contributed to the analysis, almost 70 percent were for-profit commercial companies, almost 20 percent were nonprofit commercial companies or public-private partnerships, and the remaining portion was governmental. The commercial companies and organizations ranged in size from under 100 to well over 100,000 employees.[8] Appendix A provides a more detailed breakout of the companies and organizations interviewed.

We asked each of the organizations about their cyber organizational and workforce practices, including: how their cyber-related departments were organized, which cyber functions they have in common with USAF, their strengths with respect to those cyber functions, and the enablers for achieving those strengths, such as approaches to hiring, training, and retaining skilled personnel. Appendix B provides more detail on the types of questions asked. All of our findings applied to at least 90 percent of the 15 for-profit commercial companies interviewed, with additional support from at least one nonprofit commercial company.

A company's degree of commonality with USAF was determined along four different axes: size, cyber functions performed, threat, and operational environment.

Size

As Figure 1.1 shows, few companies rival USAF in size. With nearly 330,000 members of the active component alone, and well over 450,000 personnel when including civilian and reserve component equivalents,[9] USAF is larger than all but five U.S. global corporations reported in Forbes "Global 500" lists. However, more than 50 U.S. companies exceed 100,000 employees, putting them at the same order of magnitude as USAF. We included several of these companies in our interviews. These companies were chosen since they required a large information technology (IT) infrastructure, approaching the scope of USAF's infrastructure. Additionally, large corporations are appealing targets for cyber crime or cyber espionage; therefore, these companies uniformly had significant resources invested in protecting their large IT infrastructure.

---

[7] To maintain their competitive advantage, many of these companies agreed to participate in the interviews only on the condition of anonymity. Therefore, we do not report company names in what follows.

[8] The smaller companies interviewed tended to have expertise in cybersecurity, with additional knowledge of approaches employed across their client bases.

[9] Air Force Personnel Center, "Air Force Personnel Demographics," March 31, 2014.

Figure 1.1. Few Companies Are as Large as USAF

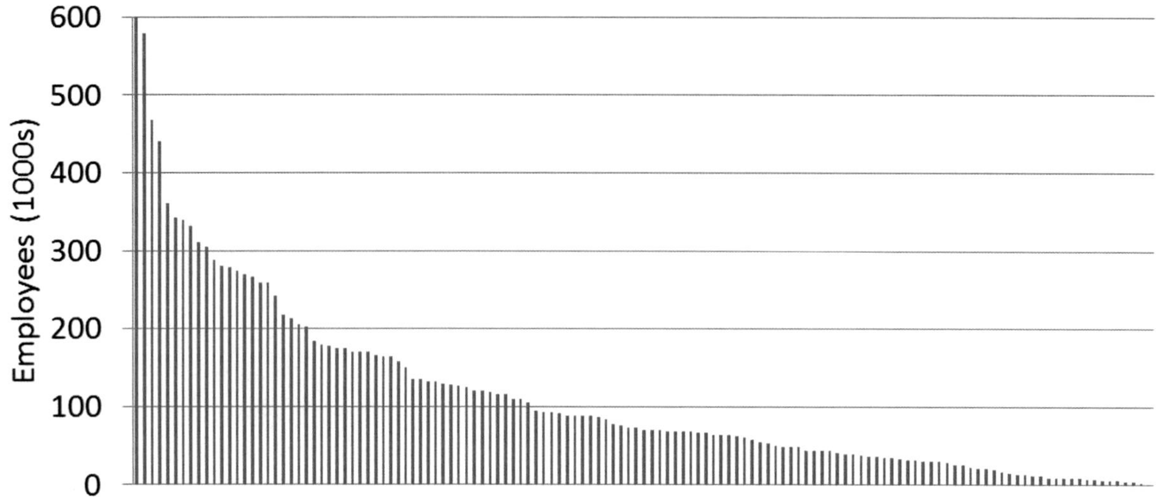

SOURCE: Fortune, "Global 500 2014," undated.

NOTE: The horizontal axis displays the top 500 companies in order of decreasing size.

Cyber Functions

USAF describes three types of cyber functions: Department of Defense information network operations (DoDIN Ops), defensive cyber operations (DCO), and offensive cyber operations (OCO).[10] Similar to DoDIN Ops, the commercial sector must design, build, configure, operate, and sustain its own networks. The commercial sector must also ensure that its networks are protected, threats are analyzed, and breaches investigated; these functions share similarities with DCO. Although the commercial sector might employ "ethical hackers" and penetration testers to test their own networks, they are legally prevented from engaging in offensive operations on outside networks. So although we can use proxies for OCO functions, there is not a perfect parallel to this USAF mission in the commercial sector. However, we interviewed four network security specialist firms that contained elements of both DCO and OCO in the services they provided. We also interviewed three organizations that specialized in forensics; however, since the role of forensics is limited to the USAF's Office of Special Investigations, their applicability to USAF as a whole was limited.

---

[10] DoDIN Ops are "operations to design, build, configure, secure, operate, maintain, and sustain Department of Defense networks to create and preserve information assurance on the Department of Defense information networks." DCO are "passive and active cyberspace operations intended to preserve the ability to utilize friendly cyberspace capabilities and protect data, networks, net-centric capabilities, and other designated systems." OCO are "operations intended to project power by the application of force in or through cyberspace." See Joint Publication 1-02, *Department of Defense Dictionary of Military and Associated Terms*, Washington, D.C.: Department of Defense, November 8, 2010, as amended through August 15, 2014.

Threat

Like the USAF, the commercial sector also faces advanced persistent threats.[11] According to a recent analysis of more than 47,000 commercial security incidents, nearly one-fifth of confirmed data breaches were attributed to state-affiliated actors (tied to China) attempting to steal intellectual property.[12] Furthermore, both USAF and the commercial sector are vulnerable to insiders. Whether malicious or negligent, insiders' actions can lead to the disclosure of sensitive information or otherwise aid outside attackers. Sixty-nine percent of recent commercial security incidents reportedly originated from either a malicious or negligent insider.[13] While the commercial sector at large faces many of the same threats as USAF, we selected well-known companies that are likely to be attacked, companies who had experience fending off and/or recovering from attacks, and computer security companies who specialize in advising about countering threats.

Operational Environment

Finally, USAF has the unique requirement for robust and secure cyber capabilities in both a home station setting and an expeditionary, deployed environment for wartime operations. The parallels with the commercial sector are obvious at home station, but there are few companies likely to experience environments that are simultaneously expeditionary *and* threatened. We interviewed three organizations that operate networks in austere deployed environments or places with a degraded communication network. However, only one of these organizations emphasized the need to operate in such an environment simultaneously with a severe cyber threat. We will return to this limitation later in the document when we describe the extent to which commercial practices are applicable to USAF.

Pervasive Practices, Plus Support from the Literature

Although we endeavored to identify best practices, the literature review and interview results clearly indicated that it is difficult to determine what constitutes *best*. In this report, therefore, we identify practices that were common across many of the commercial organizations we interviewed, across a wide range of industry sectors, and are supported by additional evidence from the literature. Due to these companies' reputation for cyber competence and, in many cases, excellence, we believe that this approach will not erroneously recommend a flawed practice for widespread adoption. However, this approach will likely overlook some highly effective practices that are either appropriate only in certain settings or are so cutting-edge as to precede widespread industry adoption.

---

[11] Mandiant, *APT1: Exposing One of China's Cyber Espionage Units*, February 2013.

[12] Verizon RISK Team, *2013 Data Breach Investigations Report*, 2013, p. 5.

[13] Verizon RISK Team, 2013, pp. 19–20.

Despite being unable to clearly define *best*, we did observe significant similarities between the practices employed by the disparate commercial companies we interviewed. We consider whether these common practices are applicable to USAF and how they could best be implemented. First, however, we describe our approach for articulating commercial cyber practices.

A Lexicon for Describing Commercial Cyber Practices

Since 2008, the Department of Defense (DoD) has defined *cyberspace* as a "global domain" consisting of the "interdependent networks of information technology infrastructures" and their "resident data," including networks like the Internet, telecommunications networks, computer systems, and "embedded processors and controllers."[14] To operate in cyberspace, USAF manages its cyber workforce in various specialties defined by officer and enlisted Air Force Specialty Codes (AFSCs), as well as Air Force civilian service occupation codes for civilian personnel. Not surprisingly, the commercial sector does not use the same terminology to describe the specific cyber functions performed by its personnel. For this reason, we employ an established lexicon—the National Initiative for Cybersecurity Education (NICE) Framework—for cyber practices that is commonly understood and sufficiently descriptive.

National Initiative for Cybersecurity Education Framework

NICE developed a framework to provide a common understanding of and lexicon for cybersecurity work.[15] The National Institute of Standards and Technology (NIST) leads the NICE initiative, in collaboration with more than 20 federal departments and agencies. Because the framework is built in collaboration with a wide spectrum of government entities and intended for use in both the public and private sectors, we adopted its taxonomy of cyber functions in our study.

The NICE Framework consists of seven categories that group together related specialty areas. The seven categories, with description of the types of specialty areas included, are:[16]

- **Securely provision**—Responsible for conceptualizing, designing, and building secure IT systems (i.e., responsible for some aspect of systems development). Specialties include: information assurance compliance, software assurance and security engineering, systems security architecture, technology research and development, systems requirements planning, test and evaluation, and systems development.

---

[14] Joint Publication 1-02, 2010.

[15] National Initiative for Cybersecurity Education, *National Cybersecurity Workforce Framework*, Washington, D.C.: Department of Commerce, 2013.

[16] National Initiative for Cybersecurity Education, 2013.

- **Operate and maintain**—Responsible for providing the support, administration, and maintenance necessary to ensure effective and efficient IT system performance and security. Specialties include: data administration, knowledge management, customer service and technical support, network services, system administration, and systems security analysis.
- **Protect and defend**—Responsible for identification, analysis, and mitigation of threats to internal IT systems or networks. Specialties include: computer network defense analysis, incident response, computer network defense infrastructure support, and vulnerability assessment and management.
- **Collect and operate**—Responsible for specialized denial and deception operations and collection of cybersecurity information that might be used to develop intelligence. Specialties include: collection operations, cyber operations planning, and cyber operations.
- **Analyze**—Responsible for highly specialized review and evaluation of incoming cybersecurity information to determine its usefulness for intelligence. Specialties include: threat analysis, exploitation analysis, all source intelligence, and targets.
- **Investigate**—Responsible for the investigation of cyber events and/or crimes of IT systems, networks, and digital evidence. Specialties include: digital forensics and investigation.
- **Oversight and development**—Responsible for providing leadership, management, direction, and/or development and advocacy so that individuals and organizations might effectively conduct cybersecurity work. Specialties include: legal advice and advocacy, strategic planning and policy development, education and training, information systems security operations oversight, and security program management oversight.

Although the NICE Framework covers a wide spectrum of jobs related to cyber, there is not a one-for-one mapping with USAF terminology. For example, DoDIN Ops includes many of the functions in *operate and maintain*, but also contain elements of *securely provision*. DCO primarily parallels *protect and defend* functions; it also contains elements of *securely provision* and, to the limited extent that they are performed in the commercial sector, *collect and operate* and *analyze* functions. Furthermore, many of the recent trends in computing are not applicable to only a single NICE functional area—technology innovations such as cloud computing, process improvements like Agile,[17] and the increasing emphasis on social media and mobile access are relevant to more than one of the functional areas. Yet the NICE Framework is still useful for this analysis because it provides sufficient detail (by describing the tasks performed and the knowledge, skills, and abilities required for each specialty) for a common lexicon, enabling a description of successful commercial practices and a comparison between USAF and the commercial sector.

---

[17] The Agile process assesses the direction of software development throughout its life cycle, with regular cadences of work, known as *sprints* or *iterations*, and every aspect of development—requirements, design, and so on—is continually revisited to ensure that a product maintains its market relevance. Agile Methodology, website, undated.

While NICE provides a common framework for describing cyber functions, this lexicon is still rather new. Thus, the organizations we interviewed still tended to use their own terminology to describe the types of cyber activities they conduct. During our interviews, we converted each organization's terminology into NICE terminology as a way of standardizing across all the organizations we interviewed. Again, Appendix B details the questions asked in our semistructured interviews and illustrates how the NICE Framework was integrated. In what follows, we focus on the first five elements of the NICE Framework.

Commercial Cyber Functions Are Often Described as IT *and* InfoSec

In conducting our interviews, we observed that organizations tended to refer to functions associated with providing cyber services as *IT* functions. This includes actions associated with the operation and maintenance of computer systems, networks, and data; requirements planning; knowledge management and in-house software and systems development;[18] computer user and network support; and software assurance. The functions described as *IT* align well with the NICE categories *operate and maintain* and parts of *securely provision.*

Organizations often used *information security* (InfoSec) to describe the functions associated with protecting and defending systems and networks; detecting, investigating, and responding to security incidents; ensuring information assurance compliance; performing security systems development; and designing system security architectures. Some organizations also described threat analysis and active defense as part of InfoSec. As such, InfoSec aligns well with the NICE categories *protect and defend*, parts of *securely provision*, and, to the limited extent that they are performed in the commercial sector, *collect and operate* and *analyze.*

Structure of the Report

The rest of this report is organized around our main findings. In each chapter, we present evidence from both the interviews and the literature review that support each finding.

Chapter Two describes our finding that there are *two* distinct sets of commercial cyber workforce practices: One set of practices dictates how IT departments assign job roles, train their workforce, and progress them through their careers; there is a different set of practices for InfoSec departments.

Chapter Three describes the evidence for our finding that IT remains a critical core function performed by a large staff, and it describes outsourcing practices.

---

[18] We observed software-development functions taking place as part of both IT and InfoSec. IT software-development capability can include web and mobile applications for use by corporate employees, as well as, to a lesser extent, according to our interviews, for customer use. InfoSec software development includes systems related to security monitoring and recovery. However, companies also reported significant development activities taking place in the business units (i.e., outside their departments responsible for IT and InfoSec) when hardware and software are part of the products the company produces and sells.

Chapter Four describes our findings related to management structures and practices for cultivating technical depth in staff and leadership.

Chapter Five focuses on hiring and retention. Here, our finding is that commercial practices favor traditional methods for recruiting and retention.

Chapter Six considers the unique constraints that USAF faces and evaluates which common commercial practices might be applicable to USAF.

Chapter Seven summarizes our findings and concludes with suggested actions for USAF.

Appendix A catalogs the characteristics of the companies and organizations interviewed.

Appendix B provides more detail on the types of questions asked during our semistructured interviews.

Appendix C is a primer on organizational design.

Appendix D details the various organizational constructs for InfoSec teams.

2. IT and InfoSec Have Different Workforce Management Practices

This chapter describes our finding that there are *two* distinct sets of commercial cyber workforce practices: one set of practices dictates how IT departments assign job roles, train their workforce, and progress them through their careers; a second set of practices also exists for InfoSec departments.

The commercial sector trains and develops IT staff in different ways from how they train and develop InfoSec staff. The fact that the commercial sector finds value in differentiating between IT and InfoSec disciplines might foreshadow similar benefits for USAF. We evaluate the extent to which that is true later in this report. First, however, we describe the basis for the commercial practices for IT and InfoSec workforce management.

Job Roles Differ Between IT and InfoSec

Based on the previous description of what constitutes IT and InfoSec, it is no surprise that job roles are different for the two disciplines. IT job roles include operating and maintaining computer systems, networks, and data; requirements planning; performing systems and software development; providing customer service and technical support; and conducting software assurance. In contrast, InfoSec job roles relate to protecting and defending systems and networks; detecting, investigating, and responding to security incidents; ensuring information assurance compliance; performing security systems development; and designing system security architectures. Table 2.1 lists some of the roles and responsibilities we encountered in the commercial sector, expressed in NICE terminology.[19]

---

[19] As shown in Table 2.1, both IT and InfoSec job functions can include software development, depending on the nature of the system under development. Furthermore, InfoSec job roles can involve assessing and ensuring the security of software developed by other corporate business units. We did not conduct an assessment of commercial practices related to software development (e.g., Agile) as this was outside our scope.

Table 2.1. Representative IT and InfoSec Job Roles

	IT	InfoSec
NICE Framework functional areas	• Operate and maintain • Securely provision	• Protect and defend • Securely provision • Collect and operate • Analyze
Job roles	• Provide tiered-level customer support • Develop and administers databases and/or data management systems • Manage and administer tools to allow the identification of, documentation of, and access to intellectual capital • Install, configure, test, operate, and maintain networks and firewalls, including hardware and software • Install, configure, troubleshoot, and maintain server configurations to ensure confidentiality, integrity, and availability • Manage accounts, firewalls, and patches • Conduct integration, testing, operations, and maintenance of systems' security • Consult with employees to gather and evaluate requirements and translate to technical solutions • Develop systems to meet requirements, following software assurance best practices	• Identify, analyze, and report events to protect information, information systems, and networks from threats • Test, implement, deploy, maintain, review, and administer infrastructure hardware and software required to manage computer network defense • Monitor network to remediate security issues • Respond to crisis situations to mitigate immediate and potential threats • Assess threats and vulnerabilities, assess risk, and develop mitigation solutions • Ensure that new IT systems meet information assurance requirements • Synthesize intelligence information • Conduct exploitation analysis • Conduct threat analysis • Develop security systems • Analyze threat information • Apply current knowledge of threat actors

SOURCES: Interviews; National Initiative for Cybersecurity Careers and Studies, *Interactive National Cybersecurity Workforce Framework*, Washington, D.C.: Department of Homeland Security, undated.

Training Differs Between IT and InfoSec

The differences in job roles are reflected in the need for different training regimens for IT and InfoSec. In our research, we observed differences in formal education (e.g., college coursework), professional training (e.g., certifications), and on-the-job training.

Formal Education

It was common practice for the companies to hire staff with bachelor's degrees, because the degree provides a strong foundation of relevant knowledge and demonstrates an ability to succeed in a professional setting. There is strong consensus about the type of education that prepares students for a career in IT, and courses of study are well established. Historically, degrees in subjects such as computer engineering, computer science, information systems, and software engineering have led to careers in IT. Since the mid-2000s, however, dedicated undergraduate degree programs in IT have proliferated.[20] A detailed report cosponsored by the Association for Computing Machinery (ACM) and the Computer Society of the Institute of Electrical and Electronics Engineers (IEEE) examines the IT field and describes key components of an IT curriculum,[21] including 33 courses that form the basis of a degree program. Such IT programs are tailored to examine "issues related to advocating for users and meeting their needs within an organization and societal context through the selection, creation, application, integration and administration of computing technologies."[22]

Several other degree programs can also lead to an IT career, including computer engineering, computer science, information systems, and software engineering. Computer engineering largely focuses on the design and implementation of systems from both a hardware and a software perspective; computer science encompasses a large discipline, including the design (with theoretical and practical perspectives) and implementation of software, new ways to use computers, and new ways to solve computation problems; the information systems program "focus[es] on the broader role of IT-enabled information utilization and business processes in a wide range of enterprises, while still maintaining [the programs'] close association with business schools;"[23] and software engineering is closely related to computer science, but with a stronger emphasis on software development.

It has been determined that none of these degree programs—IT, computer engineering, computer science, information systems, software engineering—encompass InfoSec skill sets at a curricular level.[24] An assessment of these degree programs shows a shortfall in at least one of the security topics deemed to be a minimal qualification for InfoSec.[25] For computer science in particular, an "information assurance and security" knowledge area encompasses much of what an InfoSec degree program might examine[26]; however, this knowledge area is one of 18 that a

---

[20] Master's degrees in IT have become popular enough to be ranked by *U.S. News*, and undergraduate degree programs consist of both online and traditional classroom programs. As a partial list, see the online IT degree programs at Arizona State University, Northeastern University, Pennsylvania State University, and the University of Massachusetts, as well as the classroom IT degree program at the Rochester Institute of Technology. See Joint Task Force for Computing Curricula, *Computing Curricula 2005: The Overview Report*, Cambridge, Mass.: Association for Computing Machinery and Institute of Electrical and Electronics Engineers, 2006, p. 33.

[21] Barry Lunt, Joseph J. Ekstrom, Sandra Gorka, Gregory Hislop, Reza Kamali, Eydie Lawson, Richard LeBlanc, Jacob Miller, and Han Reichgelt, *Information Technology 2008: Curriculum Guidelines for Undergraduate Degree Programs in Information Technology*, Cambridge, Mass.: Association for Computing Machinery and Institute of Electrical and Electronics Engineers, 2008.

[22] Joint Task Force for Computing Curricula, 2006, p. 9.

[23] Joint Task Force for Computing Curricula, 2006, p. 5.

[24] Joint Task Force for Computing Curricula, 2006, p. 5.

[25] Joint Task Force for Computing Curricula, 2006, p. 24.

[26] Joint Task Force on Computing Curricula, *Computer Science Curricula 2013*, Cambridge, Mass.: Association for Computing Machinery and Institute of Electrical and Electronics Engineers, 2013, p. 97.

computer science program must cover, and is therefore viewed as cursory preparation for an InfoSec career.

In fact, InfoSec curricular development is still a work in progress. The academic community has actively worked for more than a decade toward developing educational programming for InfoSec, particularly in higher education. In 1998, President Bill Clinton stated in Presidential Decision Directive 63 that "the White House . . . shall consider a series of conferences . . . that convoke academic leaders from engineering, computer science, business and law schools to review the status of education in information security and will identify changes in the curricula and resources necessary to meet the national demand for professionals in this field."[27] Along these lines, much of the academic work in InfoSec curriculum development has been funded through NIST,[28] the National Science Foundation,[29] the National Security Agency (NSA) in conjunction with the Department of Homeland Security (DHS),[30] and the Department of Education.[31] Furthermore, the National Academies recently recommended against formalizing a professional credentialing program in InfoSec until the field can be described by well-defined, stable characteristics.[32]

However, in the course of InfoSec curriculum development, the academic community has thought broadly about various issues related to educating students about InfoSec. One area of investigation is creating a common body of knowledge for InfoSec.[33] Another approach is developing hands-on exercises for students to participate in, from local to national competitions as well as university Internet-scale simulations.[34] These competitions and simulations train students in both offensive and defensive techniques. (We will return to the topic of competitions

---

[27] White House, *The Clinton Administration's Policy on Critical Infrastructure Protection: Presidential Decision Directive 63*, white paper, Washington, D.C., 1998.

[28] National Initiative for Cybersecurity Education, 2013.

[29] National Cyber Watch Center, homepage, undated.

[30] National Security Agency and Central Security Service, "National Centers of Academic Excellence in Information Assurance (IA)/Cyber Defense (CD)," posted on January 15, 2009, last modified August 20, 2014.

[31] There are numerous academic conferences, organizations, and journals addressing curriculum development for InfoSec, some within the larger context of IT education research. Conferences for InfoSec education include the Colloquium for Information Systems Security Education, the Information Security Curriculum Development Conference, and the Information Systems Education Conference. In addition, there is also an interest group within the ACM dedicated to information technology education, including InfoSec education. There are numerous journals that have regularly published information security education articles; these journals include the *Journal of Information Systems Education*, the *Information Systems Education Journal*, and *IEEE Transactions on Education*.

[32] National Research Council, *Professionalizing the Nation's Cybersecurity Workforce? Criteria for Decision-Making*, Washington, D.C.: The National Academies Press, 2013, p. 4.

[33] Marianthi Theoharidou and Dimitris Gritzalis, "Common Body of Knowledge for Information Security," *IEEE Security & Privacy* 5, no. 2 (2007).

[34] Lance J. Hoffman, Tim Rosenberg, Ronald Dodge, and Daniel Ragsdale, "Exploring a National Cybersecurity Exercise for Universities, *IEEE Security & Privacy* 3, no. 5 (2005).

and hands-on exercises as they relate to recruiting talented personnel in later chapters.) Academics also investigated how to include real-world case studies from all available sources, including industry and government, to maintain technical relevance.[35] There has also been academic work thinking beyond the technical aspects of InfoSec, including how to motivate students to think critically and creatively about InfoSec issues, as well as try to contextualize security needs to various deployment scenarios.[36]

InfoSec formal education is considerably less mature and defined than IT formal education. Therefore, we conclude that formal education needs differ for IT and InfoSec, even if the formal education programs that currently exist lag that need.

Professional Training

Given the nascent status of InfoSec formal education, it is no surprise that professional training is often used to fill the gap on specific topics of use in the workplace. In particular, certifications in the use of security principles and tools are highly sought after, both by practitioners and their employers. Here again, we point to a difference in the certifications useful for IT and those useful for InfoSec.

Certifications

We found evidence that certifications are viewed as more important for a career in InfoSec than for IT. However, even for InfoSec, holding a certification is an indicator of a staff member's basic competence but not necessarily an indicator of excellence.

A 2013 study surveyed 12,396 "qualified information security professionals" online.[37] Sixty-eight percent of respondents indicated that InfoSec certifications indicate competence, and 53 percent responded that certifications represent a higher quality of work. Interestingly, while defense-oriented government employers in the study were more likely than nongovernment employers to require InfoSec certification as a prerequisite for employment (84 percent of government employers versus 42 percent of all others), the private sector is *more* likely to view InfoSec certifications as an indicator of competency (74 percent of private sector employers versus 46 percent of public sector employers, half of whom are government defense employers).

---

[35] Sanjay Goel, Damira Pon, Peter Bloniarz, Robert Bangert-Drowns, George Berg, Vince Delio, Laura Iwan, Thomas Hurbanek, Sandoor P. Schuman, Jagdish Gangolly, Adnan Baykal, and Jon Hobbs, "Innovative Model for Information Assurance Curriculum: A Teaching Hospital," *ACM Journal on Educational Resources in Computing* 6, no. 3 (September 2006).

[36] Matt Bishop, "Teaching Context in Information Security," *ACM Journal of Educational Resources in Computing* 6, no. 3 (September 2006).

[37] Michael Suby, *The 2013 (ISC)^2 Global Information Security Workforce Study*, Mountain View, Calif.: Frost & Sullivan, 2013. This study was performed by Frost & Sullivan, in partnership with the International Information Systems Security Certification Consortium, known as (ISC)^2, and Booz Allen Hamilton.

By contrast, certification is not viewed as crucial for a career in IT. The 2013 HDI Support Center Practices and Salary Report,[38] which focused on the support center component of IT, found that only 10 percent of respondents require formal certification for their employees, and less than 25 percent think that certifications are important criteria for hiring. Of the IT occupations listed by the Bureau of Labor Statistics, only one-third mention certifications as a means of obtaining a job in those occupations;[39] however, for each of those occupations, a bachelor's degree is required by "most" employers, and certification is viewed as demonstrating "a level of competence . . . and may provide a jobseeker with a competitive advantage."[40]

To compare the different types of training provided in certification programs, we chose two representative certifications—Microsoft Certified Solutions Expert (MCSE), a well-known IT certification, and Certified Information Systems Security Professional (CISSP), a widely held InfoSec certification. These are two of some of the most popular certifications in their respective fields.[41] Table 2.2 details the training topics for each of the certifications. Aside from *security fundamentals* in MCSE and *legal*, *regulations*, *investigations*, and *compliance* in CISSP, MCSE focuses squarely on IT topics, and CISSP concentrates on InfoSec topics.

To further illuminate the role of certifications to companies, we conducted a cursory analysis of 613 job openings from one particular job website for a subset of IT and InfoSec jobs to see the types of certifications that companies are requiring.[42] As shown in Table 2.3, we find that of the IT job listings requiring a certification, like MCSE, the majority requires only an IT certification. Likewise, of the InfoSec job listings requiring a certification, like CISSP, the majority requires only an InfoSec certification.

We see that most IT jobs we analyzed (83 percent) did not require any certification. On the other hand, the majority of InfoSec jobs we analyzed (55 percent) did require a certification, and predominantly an InfoSec certification. We conjecture that this might be because InfoSec is a relatively less mature field compared with IT, and employers rely more heavily on certifications to judge competence. Some InfoSec jobs also required IT certifications (12 percent), and this could be due to a desire to guarantee some baseline understanding of IT.[43] Nevertheless, our informal analysis of a snapshot of job postings requiring certifications adds credence to our

---

38 HDI is the professional association and certification body for the technical service and support industry. HDI, *2013 Support Center Practices and Salary Report*, Colorado Springs, Colo.: HDI, 2013.

39 The occupations are computer programmer, database administrator, and network and computer systems administrator. Bureau of Labor Statistics, "Computer and Information Technology Occupations," in *Occupational Outlook Handbook*, Washington, D.C., January 8, 2014a.

40 Bureau of Labor Statistics, "Network and Computer Systems Administrators: How to Become a Network and Computer Systems Administrator," in *Occupational Outlook Handbook*, Washington, D.C., January 8, 2014b.

41 GoCertify, "Most Popular IT Certifications," web page, undated.

42 Indeed, homepage, undated. Note that this is not a comprehensive or longitudinal analysis.

43 However, there is still a sizable minority of InfoSec jobs (45 percent) that require neither type of certification.

finding that employers value skill sets that differ between IT and InfoSec, reflected in their request for different types of certifications, and that the professional training associated with these different certifications provides recipients with different preparation—either IT or InfoSec.

Our interviews echoed this finding. Although most of the senior cyber executives we interviewed did not place much value on certifications, they reported them as a valid proxy for differentiating between IT and InfoSec,[44] and some stated that certifications demonstrated that a job candidate had serious dedication to the field.

Table 2.2. Training Topics for Representative IT and InfoSec Certifications

IT (MCSE)	InfoSec (CISSP)
• Operating system fundamentals	• Access control
• Administration fundamentals	• Telecommunications and network security
• Networking fundamentals	• InfoSec governance and risk
• Security fundamentals	• Software development security
• Installing and configuring	• Cryptography
• Administering	• Security architecture and design
• Configuring advanced services	• Operations security
• Designing and implementing an infrastructure	• Business continuity and disaster recovery planning
• Implementing an advanced infrastructure	• Legal, regulation, investigations, and compliance
	• Physical (environmental) security

SOURCES: Microsoft, "Microsoft Technology Associate (MTA)," undated-c; Microsoft, "MCSA: Windows Server, 2012," undated-b; Microsoft, "MCSE: Server Infrastructure," undated-a; $(ISC)^2$, "CISSP—Certified Information Systems Security Professional," undated.

Table 2.3. Certification Requirements for IT and InfoSec Job Postings

	Certification Type (%)			
Job Type	**IT**	**InfoSec**	**Both**	**Neither**
IT	16	5	3	83
InfoSec	12	54	11	45

SOURCE: Indeed, undated.

[44] At best, organizations we interviewed considered certifications an indication of competence and a commitment to the field, but not an indication of excellence. Certifications were sometimes reported as necessary to meet government contracting requirements or used as a sorting mechanism to narrow down application pools for job postings.

On-the-Job Training

Although each company is unique in its training program, which is often proprietary, some companies have discussed their curricula in the public domain. Due to the growing interest in cybersecurity, several large defense contractors are offering variations on their internal training curriculum to DoD. For example, Northrop Grumman offers access to its Cyber Academy; Lockheed Martin offers coursework through its Center for Security Analysis; Raytheon has a course catalog that offers everything from a three-hour cyber executive course to a 22-week cyber fundamentals course; MITRE contributes training material to OpenSecurityTraining.info; and Thales has invested in a Cyber Integration & Innovation Centre to provide cybersecurity training for clients via cyber simulator time, akin to flight simulators that pilots use to maintain currency.[45] Similar to the CISSP training topics, these courses tend to focus exclusively on InfoSec issues, such as types of exploits, vulnerabilities, security tools, and penetration testing. Even in the most introductory course given at Raytheon, Cyber Fundamentals, only one topic of the 22 taught has potential overlap with IT.

These publicly available training programs confirm observations we saw from our interviews, in which on-the-job trainings for IT and InfoSec personnel have little overlap and are designed to instill distinct knowledge and skills to perform distinct sets of tasks.

Career Trajectories Differ Between IT and InfoSec

Given that IT and InfoSec have different job roles and training programs, we next assessed typical career trajectories for the two fields. Since we could not identify any academic literature on the topic, we rely on interview results in this section.

Nearly all the companies we interviewed indicated that staff transitions between IT jobs and InfoSec jobs are a rare occurrence. When such transitions do occur, they are usually in the context of an IT specialist being retrained in InfoSec. In such cases, once staffers switch fields, they tend to remain an InfoSec specialist and do not return to IT. Such instances also appeared to be judged on a case-by-case basis, depending on the individual involved, rather than companies having a blanket policy for such transitions.

Career trajectories within IT and InfoSec were somewhat different based on the size of the company. For example, some Fortune 500 companies invest in an internal formal training program for recent college graduates, and then transition them to on-the-job training. On the other hand, many of the smaller technology or security-focused companies hire personnel with

---

[45] Northrop Grumman, "Cyber Academy: Developing the Cyber Workforce," 2013; Lockheed Martin, "Lockheed Martin Center for Security Analysis (LMCSA)," undated; Raytheon, *Cyber Learning Solutions: 2012–2013 Extended Course Catalog*, 2012; MITRE, "Cybersecurity Awareness and Training," 2014; OpenSecurityTraining.info, homepage, undated; and Thales Group, "Thales Unveils Cyber Integration & Innovation Centre," October 24, 2013.

significant work experience in the specialty needed. When senior personnel were hired, they were hired based on their specialty. For example, if company hired a malware reverse engineer, that person was hired based on work experience as a malware reverse engineer. Another difference articulated in the interviews was that InfoSec staff needed longer development time to become subject-matter experts. The chief executive officer (CEO) of a cybersecurity company recently described an on-the-job training time line on the order of six to eight years until a highly skilled InfoSec professional could meet the needs of the company.[46] Figure 2.1 depicts typical career progressions for IT and InfoSec staff.

Figure 2.1. Typical Career Paths for IT and InfoSec Staff

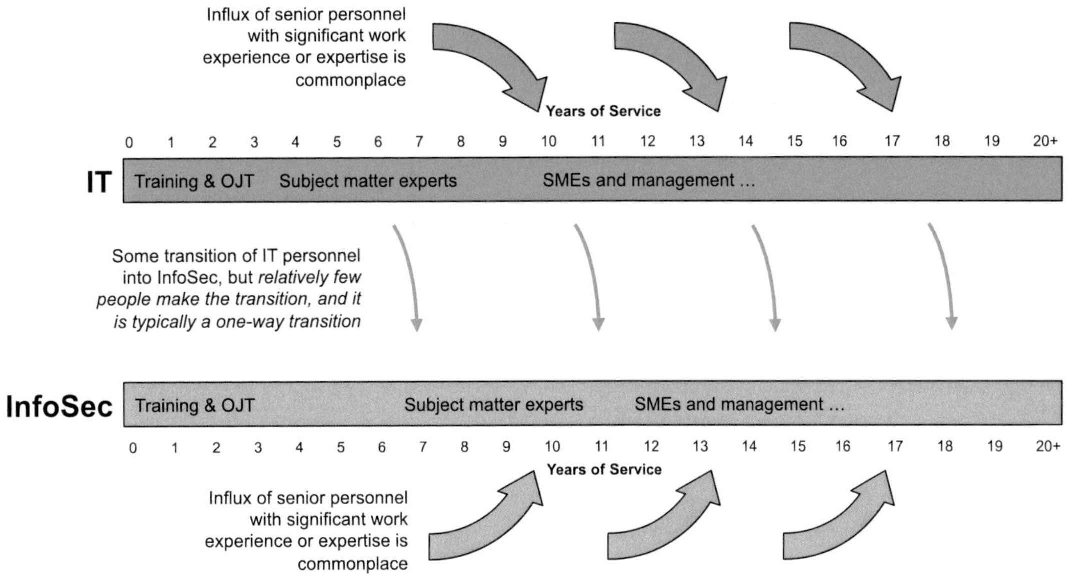

NOTE: OJT = on-the-job training; SME = subject-matter expert.

Several companies maintained formalized processes to monitor and manage career progression, not unlike USAF development teams. Such companies took steps to identify top performers on a path to higher management and provide opportunities to add professional breadth to their careers,[47] in addition to the depth they were cultivating in either IT or InfoSec. In this context, *breadth* does not refer to gaining experience in *both* IT and InfoSec; rather, *breadth* equates to gaining exposure to operations in other parts of the company. For example, a software security engineer might be based in the CISO's organization initially, but if that employee is particularly promising, he or she might be sent to work on the security aspects of a high-profile development program in one of the business units. This approach to gaining breadth serves

---

[46] Ronald Pike, "The Case for Depth in Cybersecurity Education," *ACM Inroads* 5, no. 1 (2014).

[47] However, not all employees are expected to move into the management ranks. Most of the companies we interviewed described the fact that both a technical track and a management track exist for career progression.

several purposes—it retains depth (in either IT or InfoSec) by applying the employee's specialty to a new application, it exposes the employee to the needs of the business units, and it aids retention by keeping top-performing employees challenged, working on interesting projects, and progressing in their careers.

Thus, we have seen so far that job roles, training, and career trajectories are all distinct for IT as compared with InfoSec. Therefore, it stands to reason that IT and InfoSec departments should be organized in different ways. In the next section, we bolster this intuition by describing how each field should be organized, according to guidelines from the organizational design literature.

Organizational Designs Differ Between IT and InfoSec

We observed different management structures for IT organizations and InfoSec organizations. In this area, we can point not only to the practices of the organizations we interviewed but also to the rich organizational design literature that describes best practices based on decades of observational case studies.[48] We briefly summarize the results of our analysis of how this literature applies to cyber, with a summary of the organizational design literature provided in Appendix C.

Organizational design research provides a framework for systematically analyzing and understanding how organizations function.[49] While there is no one best design for an organization under all circumstances, even in theory,[50] organizational design guidelines help answer such questions as:

- Should staff with like functions be grouped together in one department or dispersed throughout the business units?
- Should a supervisor manage many or few staff?
- How strong should the connection be between different groups?
- Should employees follow standard procedures or be granted more autonomy?

---

[48] To leverage organizational design, we examined academic management journals and books discussing organization design issues, primarily in the nonmilitary sector. We also searched past RAND reports that studied organization design issues for the U.S. military (Air Force, Army, Navy, and the Marines). See Don Snyder, Bernard Fox, Kristin F. Lynch, Raymond E. Conley, John A. Ausink, Laura Werber, William Shelton, Sarah A. Nowak, Michael R. Thirtle, and Albert A. Robbert, *Assessment of the Air Force Materiel Command Organization: Report for Congress*, Santa Monica, Calif.: RAND Corporation, RR-389-AF, 2013; Francis Fukuyama and Abram N. Shulsky, *The "Virtual Corporation" and Army Organization*, Santa Monica, Calif.: RAND Corporation, MR-863-A, 1997; Margaret C. Harrell, Harry J. Thie, Roland J. Yardley, and Maria C. Lytell, *Information Systems Technician Rating Stakeholders*, Santa Monica, Calif.: RAND Corporation, TR-1122-NAVY, 2011; Christopher Paul, Harry J. Thie, Katharine Watkins Webb, Stephanie Young, Colin P. Clarke, Susan G. Straus, Joya Laha, Christine Osowski, and Chad C. Serena, *Alert and Ready: An Organizational Design Assessment of Marine Corps Intelligence*, Santa Monica, Calif.: RAND Corporation, MG-1108-USMC, 2011.

[49] Daft, 2008, pp. 6–7.

[50] Daft, 2008, p. 26; Duncan, 1979, p. 61; Jay Galbraith, *Designing Complex Organizations*, Reading, Mass.: Addison-Wesley, 1973, p. 2.

- Should decisionmaking be centralized (at the top) or decentralized (at lower levels)?

The Operating Environment Drives Organizational Design

One of the major drivers that has been extensively studied in the academic literature is the external environment in which the organization is operating.[51] Note that the external environment is something that the organization does not have direct control over. Therefore, the organization's design must be able to effectively deal with uncertainties in the environment. The literature describes environmental *complexity* and *variability* as two of the main determinants of effective organizational design.

Environmental complexity is determined by "the number and dissimilarity of external elements, relevant to an organization's operations."[52] Well-established companies with well-known customer bases and predictable competition would fall toward the low complexity end of the spectrum—the stimuli they need to deal with are relatively few and similar. On the other hand, firms (e.g., technology start-ups) that have to identify customer bases for new, unproven products protect their intellectual property, comply with evolving government regulations, fend off fierce competition, and operate in a higher-complexity environment—the firms face numerous and diverse external stimuli.

Environmental variability is determined by how fast the environment changes over time. For example, public utilities might operate in a low-variability environment—the demand for electricity is largely predictable day to day and year to year. On the other hand, companies that need to constantly develop new products and beat the competition to market, as well as deal with the boom and bust demands of fickle customers, would experience a highly variable environment.

The organizational approach appropriate for a low complexity and variability environment is termed the *low-cost leadership* strategy, and it focuses on efficiency. Such a strategy consists of a functional departmental grouping, a large span of control, weak lateral linkages, standardization, and centralized decisionmaking, as depicted in the lower left region of Figure 2.2. The benefits of such a design are described further in Appendix C.

The organizational approach that is appropriate for a highly complex and variable environment is termed the *differentiation* strategy, and it focuses on effectiveness. Such a strategy consists of divisional, product-oriented departmental groupings, narrow spans of control, strong lateral linkages, staff creativity, and decentralized decisionmaking, as depicted in the

---

[51] Daft, 2008, pp. 138–173; Duncan 1979; Henry Mintzberg, "Organization Design: Fashion or Fit?" *Harvard Business Review* 59, no. 1 (1981). Also note that there are other elements inherent in organizations themselves that can be designed in various ways. See, for example, D. S. Pugh et al., "Dimensions of Organization Structure," *Administrative Science Quarterly* 13, no. 1 (1968). However, at the top level, the characteristics of the external environment can be used to guide organizational design.

[52] Daft, 2008, p. 145.

upper right quadrant of Figure 2.2. The benefits of such a design are described further in Appendix C.

Figure 2.2. The Environment Drives Organizational Design

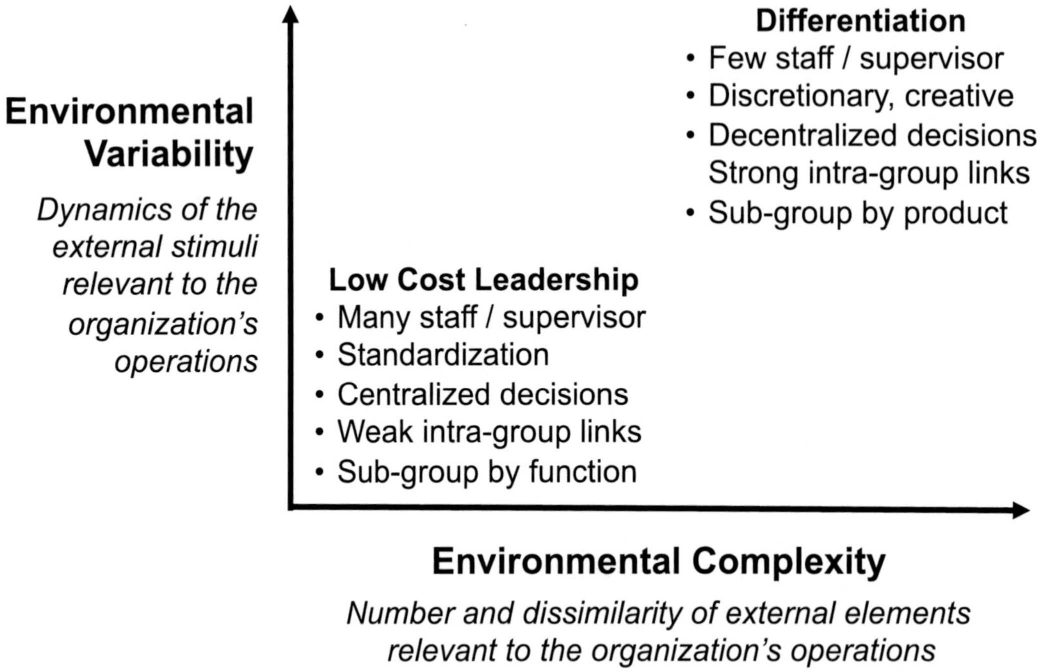

SOURCE: RAND analysis of Duncan, 1979.

Organizational Design Applied to Cyber

Although there is a significant amount of literature on organization design, we did not find any research that directly applied those principles to organizations that specialized in IT and InfoSec. In particular, we found no recommendations about organizational structure from standards-setting organizations.[53] As a result, we conducted analysis to apply the established concepts from organizational design to cyber operations.

Since environmental conditions are key drivers in organizational design, we evaluated the environmental complexity and variability for each of the NICE functional areas to determine which organization design guidelines were applicable. This evaluation is shown in Table 2.4.

We applied the guidelines about organizational design provided in Figure 2.2 to the contents of Table 2.4 to determine recommended organizational approaches for IT and InfoSec. Note, however, that we only assess environmental complexity and variability on a relative scale, and

---

[53] There is, however, guidance on IT and InfoSec governance structures, some of which refers to organizational structures to facilitate governance. For example, National Institute of Standards and Technology, *Information Security Handbook: A Guide for Managers*, Special Publication 800-100, Gaithersburg, Md., October 2006.

therefore must also evaluate the extent to which the recommended organizational approaches are applicable to IT and InfoSec. Later in this chapter, we validate our decisions for IT and InfoSec by illustrating how our observations of real-world organizations align with our assessment.

Table 2.4. Environmental Conditions Influencing Conduct of the NICE Functions

	Operate and Maintain	Securely Provision	Protect and Defend	Collect and Operate, Analyze
Environmental complexity	Driven by technology (hardware and software) that is on the system; standard procedures often used; limited possible conditions	Driven by the spectrum of technology that is available to meet user requirements; limited possible conditions, but some complex reasoning required	Driven by the variety of unknown threats that are external to the system; requires creativity	Driven by adversary behavior and possible defense mechanisms; also requires creativity
Environmental variability	Workload largely predictable by number of users and systems	System architecture not often reassessed; deliberate process	Threat changes frequently; surges common when attacked	Offense and defense changes track each other; changes frequently

IT

Our analysis suggests that environments influencing the conduct of *operate and maintain* tend to be of lesser complexity and variability. The complexity is driven by the hardware and software technology that already exists within the company's infrastructure and the nature of the company's users of these technologies. The characteristics of these users and the hardware and software baselines that support them change on deliberate timelines. The demand to perform many (although not all) tasks can be anticipated ahead of time. Furthermore, such tasks as tier 1 help desks and hardware configuration can often be clearly described and enumerated in a standardized form, having been honed to precision by previous experience. However, as the pace of technological innovation continues, the number of possible conditions under which *operate and maintain* functions must be performed will grow. Furthermore, there are surges in demand that might result in after-hour duties, but these surges can be anticipated to some extent, and they are planned for in staffing (e.g., staff on "pager duty" over the weekends). Thus, the external environment that drives *operate and maintain* is of lesser complexity in a relatively stable environment.

This is not to say that the variability or complexity of *tasks* performed in the conduct of these functions is low. Certainly, these are technical, nuanced functions. Nor is the demand completely predictable from day to day.[54] The relevant factor—and the focus of our finding—regards the variability and complexity of the *external environments* in which these functions take place. Assessment of the nature of the stimuli coming from the external environment, not the nature of the tasks, is key to identifying appropriate organizational structures, as described in the extensive literature on this subject.

Environments influencing the conduct of *securely provision* are somewhat more complex. The increasing availability of a larger number of alternatives for providing IT services indicates increased complexity in the environment.[55] The environmental dynamics are dictated by the technology refresh cycle of the business, which has historically incorporated new technologies on regular schedules. However, *securely provision* functions must increasingly adapt to the availability of game-changing technologies, indicating the somewhat higher dynamics of the environment. InfoSec also contains some elements of *securely provision*, such as information and software assurance compliance and security engineering but overall to a lesser extent than the IT community.

InfoSec

The environments influencing the conduct of *protect and defend*, *collect and operate*, and *analyze* tend to be of higher complexity and variability.

The external environments in which *protect and defend* functions occur are driven by high complexity and variability. These functions need to be able to adapt to attackers, who might act at any time and in a manner unknown to the defenders. Because of the unpredictability in the nature and timing of an attack, the demand for defenders is highly variable. Even if the timing and targets of the attacks were predictable, the types of technologies and techniques that could be brought to bear could be highly complex (novel and multifaceted), requiring creativity and critical thinking skills. Surges in capacity might also be necessary on short notice to deal with unknown threats.

Functions associated with *collect and operate* and *analyze* are also driven strongly by external factors—namely, the characteristics and actions of the targets. To be successful, these functions need to adapt to the potentially complex, layered, and unpredictable defensive measures taken by targets. This again requires creativity and critical thinking skills to outmaneuver the defense. These findings are illustrated in Figure 2.3.

Comparing Figure 2.2 and Figure 2.3 reveals standard practices for organizational design for IT and InfoSec. That is, IT should be organized according to the *low-cost leadership* strategy,

---

[54] For example, surges in user support occur—e.g., with the implementation of a new software package or procedure.

[55] Although the environments relevant to both *operate and maintain* and *securely provision* are relatively simple and static, some distinctions can be made. Though environmental variability is similar for both, the environmental complexity may tend to be higher for *securely provision* due to the larger number of variables that must be considered when planning—for example, for new network architectures and other development projects.

with many staff per supervisor; use of standardized procedures, when applicable; centralized decisionmaking at higher levels of the hierarchy; and weak intragroup linkages. Furthermore, these departments organize by function (e.g., network engineers in one department and customer service staff in another department). Likewise, InfoSec departments should be organized according to the *differentiation* strategy, with small, empowered teams of few staff per supervisor, a premium placed on ensuring creativity, strong intragroup links, and decentralized decisionmaking. Furthermore, these departments organize by product (or mission) using cross-functional teams to apply multidomain knowledge with a single purpose.

Figure 2.3. Environmental Conditions for Cyber

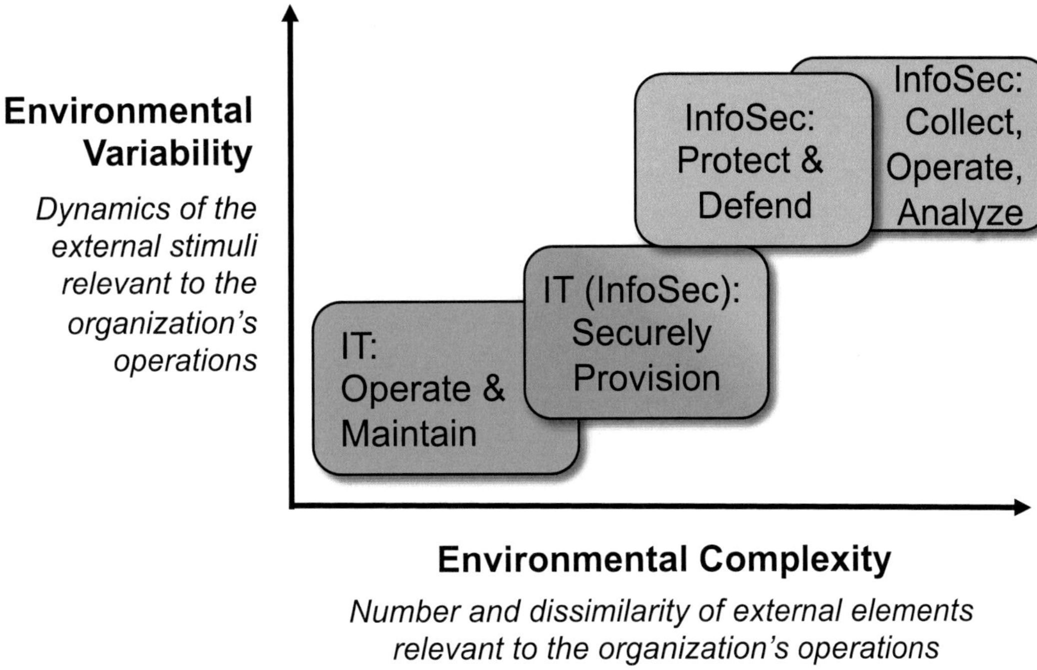

Again, this is not to say that IT organizations do not value creativity or that InfoSec organizations cannot be efficient. These broad guidelines from organizational design are, instead, a starting point for discussions of approaches to manage these organizations, based on decades of experience. The true test is whether these guidelines can apply to IT and InfoSec organizations in practice.

Observed Organizational Practices

In fact, the organizations we interviewed reported designs that echo the practices indicated by our analysis of organizational design. In what follows, we describe which elements of the low-cost leadership and differentiation organizational strategies we observed in the commercial sector, as well as whether they applied to IT or InfoSec organizations.

Many Staff per Supervisor in IT, Few Staff per Supervisor in InfoSec

From our interviews, we found that IT tends to have a large number of employees per supervisor, and InfoSec tends to have a small number of employees per supervisor, consistent with Figure 2.4. In particular, at the same level of hierarchy, we observed that IT managers supervised up to 19 times the number of staff as InfoSec managers. For example, one company partitioned its 4,000 IT employees into five hierarchical suborganizations, whereas its InfoSec group, totaling 80, was organized in small teams of five employees, including one supervisor. One company mentioned InfoSec teams sized to "two large pizzas," referring to the belief that small teams coming together over food to brainstorm ideas and solve problems were more effective than larger teams.

Figure 2.4. Organizational Design for Cyber

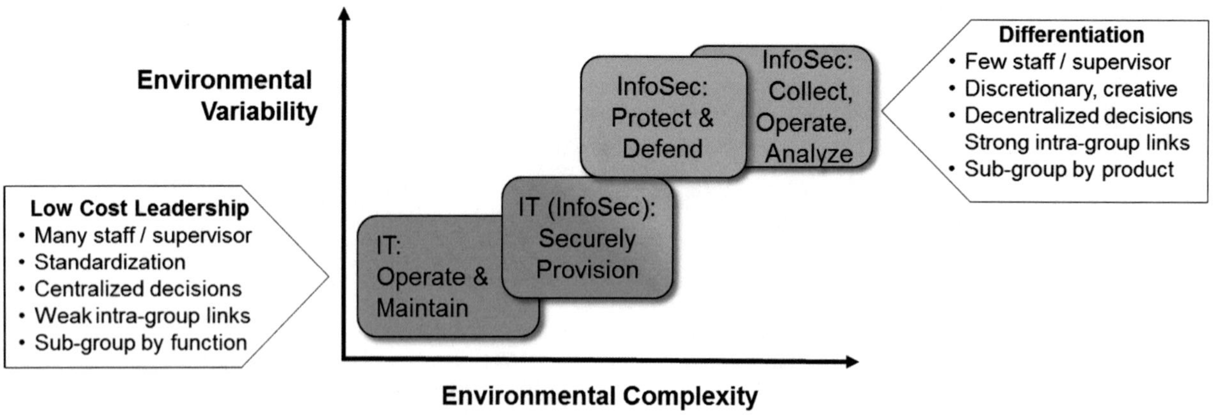

Standardization Is Emphasized in IT; Creativity Is Emphasized in InfoSec

We heard in many interviews that there is a strong emphasis on creativity and critical thinking skills in InfoSec. Such autonomy is required because of the inherent unpredictability in performing many tasks and, therefore, the lack of standardized processes on which to base InfoSec work.

Our interviews stressed, however, that this focus on creativity in InfoSec is not an indication that InfoSec personnel are superior to IT personnel. We heard repeatedly that InfoSec was not viewed as the "A team" to IT's "B team." Rather, personnel tend to self-select into one field or the other because they identify with and enjoy its style of work. Likewise, companies do not view a progression from IT into InfoSec as a typical career path. This is at least partly because the approaches to excelling in the two fields are so dissimilar—a focus on planning, developing, vetting, and adhering to established procedures in IT, contrasted with an affinity for working with higher degrees of uncertainty and disorder in InfoSec.

Decisionmaking Is Centralized for IT, Decentralized for InfoSec

Finally, decisionmaking tend to be more centralized for IT organizations headed by the CIO, whereas decisionmaking tend to be more decentralized for InfoSec organizations, where nimble teams are empowered to make decisions. In a large industrial manufacturer, the InfoSec group included a Computer Emergency Response Team (CERT) of 40 personnel. Within the CERT, those with significant InfoSec skills and experience mentored less experienced personnel through real-time operational cyber monitoring and response. Such a delegation of decision authority extended more generally to day-to-day operations and was formalized in job roles.

Linkages Between IT and InfoSec Are Common

Despite being organizationally separate, many companies discussed the importance of IT and InfoSec groups working closely to ensure seamless and effective operations. For example, one company discussed its vulnerability management process, where the InfoSec group scanned for vulnerabilities, and the IT group applied the patches. Similarly, we observed strong lateral connections in financial companies, where coordinated efforts between the IT, InfoSec, and fraud units were required to commence legal proceedings.

These lateral relationships are encouraged both formally and informally. Committees and working groups are formalized to foster relationships between different divisions. Several companies discussed having a working group or council devoted explicitly to cyber security, led by InfoSec representatives, but with representatives from many different units.

A large financial organization we interviewed used both formal and informal knowledge-sharing processes to keep IT and InfoSec aware of each other's activities. These linkages occurred at the executive levels between the CIO and CISO and also between working groups. We will describe these linkages further in the discussion of liaisons.

Suborganizations Are Aligned by Function for IT, by Product for InfoSec

Our interviews revealed that within a corporate IT organization, people were divided into subgroups by function. For example, database administrators form one division, and network services staff form a different division, both within IT. InfoSec organizations, on the other hand, formed subgroups by constructing cross-functional teams assigned to a specific mission area. For example, a security incident response group and a vulnerability assessment group would be two different mission-oriented suborganizations of a corporate InfoSec unit, each with staff with varied functional backgrounds necessary to tackle the assigned mission. Appendix D contains a description of InfoSec organizations that illustrates this point.

Consolidated Corporate Organizations, Headed by CIO and CISO

As information technologies continue to evolve, companies adapt their approaches to where to house various IT and InfoSec functions within corporate organizations. Companies continually weigh the costs and benefits to determine if staff with like functions (e.g., IT) should be grouped *together in one unit* (e.g., a consolidated IT department, reporting to a single CIO) or *dispersed throughout the company* (e.g., smaller IT groups within every business unit). Although commercial practices currently favor consolidated organizations at the corporate level for both IT and InfoSec, the calculus is slightly different in each case.

Consolidated Corporate-Level IT Organizations

Both our interviews and the literature indicate that decisions about IT consolidation are cyclical. A forward-leaning business unit adopting a new technology (e.g., cloud computing or mobile capability) before the rest of the company tends to provide the requirements analysis, architecting, and operations and maintenance functions, commonly associated with IT, within that business unit. However, as the technology matures and is adopted widely throughout the company, it becomes more efficient to consolidate provision of those IT functions. In fact, since 1990, the literature has proclaimed the benefits of consolidating IT staff for systems widely used throughout the company.[56] Several of the companies we interviewed described a process of evolving over the past decade: from many smaller IT organizations within diverse business units to a single consolidated IT organization. The reasons for doing so were usually tied to reducing costs through increased efficiency—a leaner IT staff with smaller footprints for touch maintenance in satellite offices. Consolidated management also facilitates enterprisewide assessments and a better understanding of the costs and benefits of various IT investments.

A prerequisite for such consolidation is the standardization of IT products across the company (e.g., common enterprise software and hardware) to limit the "special cases" to only those units with a crucial business need for uniqueness. However, even in the cases where some business units maintained nonstandard hardware or software, consolidated IT departments are increasingly taking responsibility for operations and maintenance of these nonstandard systems. For example, one large manufacturing firm described the fact that some of its business units need supervisory control and data acquisition (SCADA) systems to run the units' industrial processes, but even these niche systems were under the purview of the consolidated IT department, which employs specialists well versed in SCADA systems and the manner in which the units employ them.

We interviewed two multinational conglomerates, operating subsidiaries in multiple diverse industry sectors in several countries. For these conglomerates, there is a refinement to the consolidation practices described above. Because of the fact that any two of their subsidiaries might be regulated by different national governments (e.g., the United States compared with the United Kingdom) or functional bodies (e.g., DoD compared with the U.S. Department of Health and Human Services), there are sound legal reasons for keeping the subsidiaries' network infrastructures separate. Therefore, while any single subsidiary consolidates its IT function,

---

[56] Ernest M. von Simon, "The 'Centrally Decentralized' IS Organization," *Harvard Business Review* (July–August 1990).

regulatory regimes preclude consolidating *across* the entire multinational conglomerate. That is, these organizations described a need to maintain different networks, different CIOs, and different IT departments for subsidiaries that are sufficiently different. One of these conglomerates described a federated approach to managing across subsidiaries, wherein the CIO of each subsidiary ran the day-to-day operation but received longer-term policy guidance from the CIO of the parent company.

Furthermore, even smaller, less complex companies sometimes maintained a CIO (in name) within each business unit, but this position was part of the corporate consolidated IT organization—i.e., a liaison position reporting to the corporate CIO, not the business unit leadership. This liaison role was described as crucial to making a consolidated IT organization responsive to the needs of the business units. Serving as the lateral tie between a business unit and the IT organization, liaisons are collocated with the business units, and they collaborate with the staff and leadership of these units to make sure their needs are represented to the consolidated IT organization. Finally, recall that a consolidated chain of command should not be interpreted to mean that the department is centrally located. In fact, all the large organizations we interviewed described their consolidated IT department having personnel on-site at all operating locations.

Corporate-Level InfoSec Organizations

Whereas IT consolidation is typically driven by cost savings, alignment of InfoSec in a single consolidated organization is necessitated by the mission itself. That is, to *protect* the whole network, InfoSec staff must have *visibility* across the whole network. Therefore, as many of our interviews described, consolidated corporatewide management of InfoSec, under the authority of a CISO who reports to corporate leadership, ensures visibility across business units and allows the deployment of security protocols to monitor and protect the entire network.

Several companies reported that all their security functions (InfoSec and such functions as physical security and fraud prevention) have direct reporting lines to the companies' senior executives, other than the CIO (e.g., a chief risk officer, a chief administrative officer, or a chief financial officer). These companies reported that this allowed for greater independence of security advocates within the organizational structure. In fact, there is an ongoing debate in the professional trade literature as to whether the CISO should report to the CIO or another member of the "C suite" to maintain sufficient focus on security.[57] The companies we interviewed that had a CISO reporting to someone other than a CIO also mentioned that the InfoSec function was tightly integrated into the overall corporate risk management process, such that decisions about security posture and investments were made from a business risk perspective, not an IT perspective. In addition to implementing risk assessments at the enterprise level, the consolidated InfoSec management allows risk analysis to be applied at the employee level. Several companies

---

[57] See, for example, Antone Gonsalves, "Target Top Security Officer Reporting to CIO Seen as Mistake," *CSO Online*, June 13, 2014.

described processes for assigning a risk score to each employee, then monitoring employees based on their risk scores. An employee's risk score was often dependent on what sensitive data they have access to, what privileges they have on the network, and their user behavior on the network. Despite the strong connection between InfoSec and risk, Ernst & Young, among other consulting firms, has noted that only 5 percent of the companies in its sample set have InfoSec staff reporting to the chief risk officer, the person most responsible for managing a company's risk profile.[58] Ernst & Young claims that such a connection is "critical when it comes to selecting the right tools, processes and methods to monitor threats, gauge performance and identify coverage gaps."

We also observed a few instances of another organizational structure, where InfoSec reported to a head of IT, which in turn reported to the CIO. This was observed in organizations where InfoSec capabilities were relatively immature and where outsourcing was heavily utilized. It is possible this structure results in organizations that do not have strong internal InfoSec capabilities. As these organizations mature their InfoSec capabilities in-house, they might move toward a more fully separated IT and InfoSec organizational structure, with distinct reporting lines.

Researchers at the Software Engineering Institute at Carnegie Mellon University considered the organizational model of a Computer Security Incident Response Team (CSIRT) as it relates to the corporation as a whole,[59] recommending that large corporations have a CSIRT with a consolidated staff devoted to monitoring incidents across the organization and recommending security solutions. Additionally, the researchers recommend that the CSIRT have distributed team members at each business unit or remote site. According to the researchers, "The strengths of this combined model are that it provides a CSIRT composed of a stable core of professionals along with a network of affiliated members in the operating units. The centralized members provide the stability, expertise, and permanent infrastructure, while the distributed members provide the operational knowledge and expertise, along with established connections to the business units at the local levels." This distributed membership is similar in function to the business liaisons discussed earlier.

The business literature bolsters our observations that consolidation is prevalent in the commercial sector. A consolidated security organization "will provide greater consistency, influence, and control," and it supports the strong ties between InfoSec and risk management. The literature does allow that there might be some exceptions for centralization, but that is

---

[58] Ernst & Young, *Fighting to Close the Gap: Global Information Security Survey 2012*, November 2012.

[59] Georgia Killcrece, Klaus-Peter Kossakowski, Robin Ruefle, and Mark Zajicek, *Organizational Models for Computer Security Incident Response Teams (CSIRTs)*, Pittsburgh, Pa.: Software Engineering Institute, Carnegie Mellon University, 2003.

usually only reserved for "organizations with extremely autonomous business units that have very different security needs."[60]

Figure 2.5 shows this typical organizational structure for corporate-level IT and InfoSec departments, consistent with our interviews.

Figure 2.5. Consolidation of IT Under CIO and InfoSec Under CISO

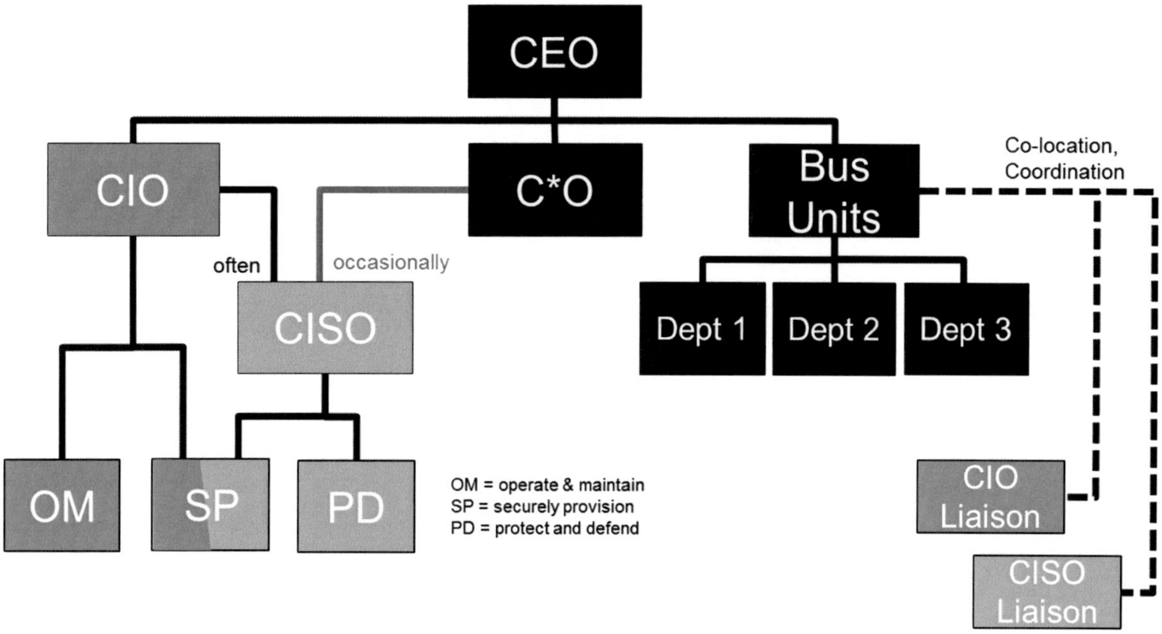

NOTE: C*O refers to any other chief officer (e.g., chief financial officer).

In the figure, IT personnel conducting *operate and maintain* functions report to the CIO, and InfoSec personnel conducting *protect and defend* functions report to the CISO. Staff performing *securely provision* functions aligned with IT report to the CIO; staff performing *securely provision* functions aligned with InfoSec report to the CISO. In many cases the CISO then reports to the CIO, but occasionally the CISO reports to a different C-level officer, such as the chief financial officer (CFO) or the chief risk officer (CRO). The liaison functions in the business units might represent single people, or large staffs, depending on the size of the business unit and its needs for IT and InfoSec support. But note that the chain of command (represented by color) is to the CIO or CISO, not to the business unit.

Finally, recall that a consolidated chain of command should not be interpreted to mean that the department is centrally located. In fact, while most companies had cybersecurity operations centers reporting to the CISO, several maintained facilities worldwide that allowed them to

---

[60] Khalid Kark and Rachel A. Dines, *Security Organization 2.0: Building A Robust Security Organization*, Cambridge, Mass.: Forrester Research Inc., 2010.

operate around the clock ("following the sun"). Thus, the consolidated InfoSec organization maintains personnel around the globe and with business units as needed to perform the mission.

Summary

In this chapter, we have explained the basis for our finding that IT and InfoSec should be organized in different ways—organizational design research implies this to be the case, and commercial practices corroborate the theory. We see that IT organizations should feature functionally aligned subgroups with many staff per supervisor, employment of standardized processes, and decisionmaking held at higher levels. InfoSec organizations, on the other hand, should feature product-oriented (or mission-oriented) subgroups, employing creativity and retaining decisionmaking at lower levels, with few staff per supervisor.

The commercial sector organizes, trains, and develops IT staff members through their careers differently from how it organizes, trains, and develops InfoSec staff members through their careers. Traditionally, it is those same factors—organization, training, and development—that have influenced whether USAF should split or combine AFSCs.

Although different, the commercial sector repeatedly emphasized that both fields are required for a company to efficiently and effectively operate their network. In fact, both the literature and our interviews repeat a common comparison—the diversity of roles found in cyber is similar to the diversity found in the medical field. That is, the medical field certainly requires practitioners with specialties (e.g., cardiology, pediatrics) but it also requires hospital administrators and medical equipment providers.[61] All of them are essential for the effective operation of a hospital, yet they are dramatically different in job roles, training, career progression, and organizational strategies. Likewise, IT and InfoSec are both critical for businesses to effectively leverage cyberspace—yet they differ with respect to job roles, training, career progression, and organizational strategy.

---

[61] Ronald C. Dodge, Costis Toregas, and Lance Hoffman, "Cybersecurity Workforce Development Directions," in *Proceedings of the Sixth International Symposium on Human Aspects of Information Security & Assurance: HAISA 2012*, ed. Nathan Clarke and Steven Furnell, Plymouth, UK: University of Plymouth, 2012.

3. IT Is a Critical Core Function Performed by a Large Staff

Companies rely heavily on their cyber systems, and most of the ones we interviewed considered IT a core competence that must be retained and resourced internally. Corporations recognize the importance of IT functionality and, as such, significant corporate resources are expended on IT capabilities and personnel.

At the same time, corporate recognition of the need for InfoSec is at an all-time high. Most of the companies we interviewed described learning from security challenges over the past decade, and then making organizational and investment changes as a result of the increased threats they face and the need to safeguard their intellectual property and the trust of their customers. Yet despite the growing value of InfoSec, the vast majority of corporate cyber expenditures are in IT.

On Average, 95 Percent of Cyber Workforce Is in IT and 5 Percent Is in InfoSec

We found striking similarities across the companies we interviewed regarding the relative allocation of IT and InfoSec personnel. On average, companies maintained approximately 20 times more IT personnel than InfoSec personnel. Stated differently, approximately 95 percent of a company's cyber personnel are devoted to IT, and only 5 percent are aligned with InfoSec.

The consistency of this observation was surprising. To validate the finding, we consulted data from the Bureau of Labor Statistics. The *Occupational Outlook Handbook* is updated and released biennially and includes information about the nature of work, working conditions, training and education, earnings, and job outlooks for hundreds of different occupations. Table 3.1 describes five different occupations that compose the bulk of the IT and InfoSec workforce. From left to right, the table lists the occupation name, a brief job description, the median pay, the number of U.S. jobs in that occupation, the expected rate of growth over the next ten years, and association with IT or InfoSec.

Based on these data, the total IT labor population of 1.35 million people account for about 95 percent of the total cyber occupation field, whereas InfoSec personnel, which is only about 75,000 people, account for the remaining 5 percent—the same relative level of effort we observed in our interviews.

Although the labor population confirms the data we collected in interviews, we do not presume this relative level of effort will hold going into the future. Looking at the projected growth rate for each occupation, the job market for InfoSec analysts is expected to grow by almost 40 percent from 2012 to 2022, whereas the traditionally IT occupations are relatively steady, with growth projected between 10 and 20 percent. Despite this rapid projected growth in

InfoSec, this translates into a *future* workforce proportion of 94 percent IT and 6 percent InfoSec through 2022.

Table 3.1. U.S. Labor Populations in IT and InfoSec

Bureau of Labor Statistics Occupation	Job Summary	2012 Median Pay ($)	Number of Jobs, 2012	Job Growth Outlook, 2012–2022 (%)	Career Field
InfoSec analysts	InfoSec analysts plan and carry out security measures to protect an organization's computer networks and systems. Their responsibilities are continually expanding as the number of cyberattacks increase.	86,170	75,100	37	InfoSec
Computer support specialists	Computer support specialists provide help and advice to people and organizations using computer software or equipment. Some, called *computer network support specialists*, support IT employees within their organization. Others, called *computer user support specialists*, assist non-IT users who are having computer problems.	48,900	722,400	17	IT
Network and computer systems administrators	Computer networks are critical parts of almost every organization. Network and computer systems administrators are responsible for the day-to-day operation of these networks.	72,560	366,400	12	IT
Database administrators	Database administrators use specialized software to store and organize data, such as financial information and customer shipping records. They make sure that data are available to users and are secure from unauthorized access.	77,080	118,700	15	IT
Computer network architects	Computer network architects design and build data communication networks, including local area networks, wide area networks, and intranets. These networks range from a small connection between two offices to a multinational series of globally distributed communications systems.	91,000	143,400	15	IT

SOURCE: Bureau of Labor Statistics, *Occupational Outlook Handbook*, Washington, D.C., January 8, 2014c.

Finally, while we observed striking conformity to these percentages in our interviews, we noted a few exceptions. The cybersecurity companies we interviewed were obviously primarily composed of InfoSec personnel. But even companies with a broad business base (more similar to USAF) exhibited some outliers. One such company spun off some of its in-house InfoSec capability into product lines for sale to clients. This company had invested heavily in its InfoSec capability and had a higher percentage of InfoSec staff (90 percent IT to 10 percent InfoSec). Another outlying case was a government department that reported 97 percent IT and only 3 percent InfoSec; in this instance, the department kept only two cyber operations completely in-house (legal and policy) and outsourced much of the remaining capabilities. Since outsourcing was not common in the other organizations we interviewed (discussed below), this result represents a different approach that uses a much smaller in-house staff.

IT Workforce Size Depends on Industry Sector and Company Size

The number of employees each IT professional supports is highly variable and dependent on both the type of industry and the company size. A survey found that, by industry, "financial services firms have the fewest employees supported by each IT professional . . . while manufacturing and government/education/nonprofit organizations have the most employees supported by each IT professional."[62]

Table 3.2 shows the ratio of IT staff to employees as a function of company size, according to a survey of 103 organizations conducted by people3, Mercer Human Resource Consulting, and the Information Technology Association of America (ITAA) in 2003. Of note, larger companies are able to reap economies of scale and employ fewer IT personnel relative to the number of employees.

Table 3.2. Ratio of IT Staff to Employees

Number of Employees	25th Percentile	50th Percentile	75th Percentile	Organization Count
Fewer than 500	1:8	1:18	1:34	16
500 to 999	1:14	1:25	1:40	14
1,000 to 4,999	1:11	1:23	1:45	38
5,000 to 9,999	1:10	1:25	1:53	15
10,000 or more	1:23	1:40	1:112	20

SOURCE: Workforce.com, "Ratio of IT Staff to Employees," February 6, 2003.

Additionally, the survey found that manufacturing companies have relatively few IT staff per employee (as low as one IT staff member for every 112 employees), whereas data-intensive companies have more IT staff per employee (as high as one IT staff member for every 11 employees). This is consistent with our interviews. For companies that were as similarly diverse as USAF, one would conservatively expect about 25 employees per IT professional.

As mentioned above, few of the organizations we interviewed relied extensively on outsourcing, preferring to keep IT and InfoSec an in-house capability. We explore this topic further in the next section.

---

[62] Business Editors, *Survey Indicates No "One Size Fits All" Solution to IT Structures and Staffing; Joint Study Released by people3, Mercer Human Resource Consulting and ITAA*, Business Wire, February 3, 2003.

Commercial Practices Demonstrate a Cautious Approach to Outsourcing

The U.S. government constantly struggles to attract and retain technical expertise that is also in demand in the private sector; consequently, outsourcing is of perennial interest to DoD.[63] Outsourcing is attractive because of the hope of hiring experts with years of experience without having to invest the time and money to develop them internally, and because of the potential for driving down costs and gaining efficiencies through economies of scale.

Outsourcing of IT and InfoSec Is Limited

Many of the companies with whom we spoke had limited the extent to which they outsourced IT or InfoSec functions. IT outsourcing was commonly pursued for tier 1 help desk (i.e., the initial support level responsible for basic customer issues) or desktop services, because these functions were not considered part of the company's core competences. However, other IT functions (e.g., data administration, knowledge management, network services, system administration, systems security analysis, system design, requirements analysis, and user account management) were considered critical core capabilities that the company wanted to manage internally. Outsourcing in InfoSec was limited to highly specialized skills, such as the penetration testing of applications or third-party verification of internal security processes and procedures.

However, we heard from senior cyber executives at two organizations that subcontracted a considerable portion of their IT and InfoSec capabilities—neither organization was a private company. In one case, the cyber leadership conceded that the decision to outsource was a mistake, as there was an insufficient number of technical personnel internal to the organization to provide adequate oversight over the contract. Consequently, the contract was poorly written and poorly executed. In the other case, the organization outsourced network design, IT support, network defense, incident response, threat analysis, forensics, active defenses, policy setting, and legal advice. The interviewee cited the decision to outsource as driven both by cost and difficulties in attaining the necessary level of skill in an in-house capacity. In contrast to the previous case, this organization maintained an appropriate level of up-to-date expertise internal to the organization to monitor the outsource relationships successfully.

The literature supports the commercial sector's limited use of outsourcing. In particular, the literature recommends that outsourcing only be considered once a company has matured its own internal processes and concluded that an outside company can more effectively or efficiently implement them,[64] and that a company should hesitate to outsource something that is considered

---

[63] U.S. General Accounting Office, *Information Technology: DOD Needs to Leverage Lessons Learned from Its Outsourcing Projects*, Washington, D.C., 2003; Valerie Bailey Grasso, *Defense Outsourcing: The OMB Circular A-76 Policy*, Washington, D.C.: Congressional Research Service, 2005; U.S. General Accounting Office, *Information Technology: DOD Needs to Ensure That Navy Marine Corps Intranet Program Is Meeting Goals and Satisfying Customers*, Washington, D.C., 2006.

[64] Kark and Dines, 2010.

a core competence. According to one report, "Outsourcing can provide a shortcut to a more competitive product, but it typically contributes little to building the people-embodied skills that are needed to sustain product leadership."[65]

Strong Internal IT Capabilities Guide Successful IT Outsourcing

The cost savings from IT outsourcing should be broadly accounted for beyond just IT operational costs. Furthermore, IT outsourcing cannot substitute for internal IT investments.[66] According to the business literature, "You have to know your operational environment well and ensure that adequate process maturity and monitoring exists before you can even think about handing it over to someone else."[67]

To take full advantage of outsourcing, the company should have an internal capability to understand and supervise the tasks that are being outsourced and ensure that the services being delivered are consistent with and fulfill the business needs. Generally, this objective cannot be achieved without investment in internal IT and InfoSec capabilities.

A lesser-recognized *benefit* of outsourcing is knowledge transfer from highly specialized service providers, like penetration testers. However, such transfer requires a client organization to be willing and able to absorb such knowledge.[68] By working with service providers that have specialized knowledge, the client organization can learn from them. The amount of knowledge that the client organization can absorb will depend on the organic capabilities of the client. Thus, the more capable the client is, the more it can absorb from the vendor. Several of the companies we interviewed indicated that they had taken this approach to InfoSec—initially relying on the services of cybersecurity firms, learning from them, and ultimately building up their own in-house InfoSec capability.

Conflict of Interest Is a Concern for Outsourcing InfoSec

Outsourcing InfoSec should be analyzed in terms of performing at least two disparate functions: detection and prevention of security breaches.[69] There are various models for contracting with managed service security providers (MSSPs). One model is that a single MSSP is contracted for providing both detection and prevention. The largest criticism of such a model is the emergence

---

[65] C. K. Prahalad and Gary Hamel, "The Core Competence of the Corporation," *Harvard Business Review* (May–June 1990).

[66] Kunsoo Han and Sunil Mithas, "IT Outsourcing and Non-IT Operating Costs: An Empirical Investigation," *MIS Quarterly* 37, no. 1, 2013.

[67] Kark and Dines, 2010.

[68] Young Bong Chang and Vijay Gurbaxani, "IT Outsourcing, Knowledge Transfer, and Firm Productivity: An Empirical Analysis," *MIS Quarterly* 36, no. 4 (2012).

[69] Asunur Cezar, Huseyin Cavusoglu, and Srinivasan Raghunathan, "Outsourcing Information Security: Contracting Issues and Security Implications," *Management Science* 60, no. 3 (2014).

of a conflict of interest, since MSSPs are less likely to report security breaches for fear of penalties by their clients. Another model is to have two separate MSSPs (one per function); however, this can introduce inefficiencies, since combining the two function leads to process synergies that are not easily realizable if they are performed by separate organizations. An alternative proposal is to use a better-suited incentive structure, following a "carrots and sticks" approach by using a single MSSP with a contract that rewards revelations of security breaches and penalizes the MSSP if it is found responsible.[70] It should be noted that this alternative scheme was analyzed with a game theoretic and economic utility model instead of an empirical evaluation of actual firms and MSSPs engaged in such a contract.

Decision Processes and Organizational Self-Evaluation Are Required

Outsourcing decisions in governmental organizations could be considered inherently different from the private sector due to political, privacy, or national security concerns and can require entirely different decision criteria and processes.[71] Traditional outsourcing decision processes emphasize the potential efficiency benefits but lack a structured method to make those decisions based on strategic objectives. First, management must define strategic objectives that drive the decision to outsource. Examples of such objectives include focusing on the core business, leveraging external subject-matter experts, minimizing costs, maximizing workforce availability, and reducing labor-intensive processes.

Also, an organization should distinguish between outsourcing a function that is presently embodied within the organization and simply contracting for services that were never performed within the organization. This distinction is not purely semantic and would lead to a more accurate cost-benefit analysis that accounts for the entire life cycle of developing a skilled workforce from a defined baseline, which might range from no previous experience to some mix of experiences.

Furthermore, outsourcing parts or whole processes can lead to institutional loss of skill, which can diminish the ability to manage delivery—from defining requirements to executing the outsourced service. Such loss of institutional skill must be accounted for in the cost-benefit analysis, as it might affect the procurement terms in a manner that can mitigate loss of institutional knowledge. The possibility of contract termination further necessitates contingency plans that account for retraining time and costs to provide the outsourced function internally or through an alternative external source that might require acculturation. The cost of skeleton staff able to maintain adequate oversight is often overlooked.

---

[70] Cezar, Cavusoglu, and Raghunathan, 2014.

[71] J. Stark, M. Arlt, and D. H. T. Walker, "Outsourcing Decisions and Models—Some Practical Considerations for Large Organizations," in *International Conference on Global Software Engineering, 2006: ICGSE '06*, Los Alamitos, Calif.: IEEE Computer Society Press, 2006.

Although the marketplaces of IT and InfoSec providers are of different maturity, there is enough competition to offer organizations potential gains in more efficiently allocating resources to their core businesses, thus making the decision to outsource either of these functions a critical one. However, the previous discussion, based on our review of outsourcing literature, indicates that properly extracting these efficiency gains will still require retaining personnel with considerable expertise to manage these outsourced functions and close coordination to ensure that these service providers are aligned with the needs of the organization. When organizations conduct a cost-benefit analysis that includes these factors for oversight and coordination, along with the uncertainty in any such decision, they might be more likely to outsource commoditized functions within IT and InfoSec, such as help desk and penetration testing, rather than more-complicated functions, such as systems security analysis and requirements analysis, which also require greater coordination and alignment with the organizations' needs and evolution.

4. Technical Leadership Is Valued and Cultivated

We found support for technical depth and currency from the commercial sector and the academic literature. *Technical depth* refers to expertise in a single topic, such as reverse malware engineer, and *technical currency* ensures that such an engineer is familiar with the most-recent tools and techniques in use for his or her expertise. We repeatedly heard in our interviews that companies value depth of expertise in staff members. These statements are corroborated by the practices they put in place that facilitate depth—e.g., career management practices that keep a staff member rooted in one field. However, our finding is not just that staff must maintain technical depth but that leaders must as well.

Indeed, the need for technical leadership in the cyber domain is often discussed in the military literature. For example, according to the director of the Cyber Security Research Center at West Point: "Technical competency is *the* fundamental requirement for a leader in cyberspace. . . . [O]nce technical literacy is gained, it must be maintained."[72] Furthermore, cyberspace is often compared with other longer-established fields, especially with respect to maintaining currency. For example, USAF Maj. Gen. Ronnie D. Hawkins, Jr., is quoted as saying: "None of us would get on an aircraft . . . with the knowledge that the pilot and everybody on that aircraft had not been certified and also recertified," implying that operating in cyber should be no different.[73] Others comment that medicine—a field that evolves quickly, like cyber—requires practitioners to keep current on new treatments.[74]

Management Must Keep Up with the Pace of Technology

It was common practice among the companies interviewed that the management in both IT and InfoSec maintained technical expertise. Technology skills, such as programming and scripting, or knowledge of hardware, are highly perishable, given the rapid pace of technological change.[75] Part of the rationale for maintaining technical currency is that the field itself is constantly evolving, and managers must keep pace with the technology they and their staffs employ to manage effectively. Otherwise, managers will quickly find that their skills atrophy, that they

---

[72] Gregory Conti and David Raymond, "Leadership of Cyber Warriors: Enduring Principles and New Directions," *Small Wars Journal*, July 11, 2011.

[73] Amy McCullough, "Cyber Futures," *Air Force Magazine*, June 2011.

[74] Booz Allen Hamilton, *Cyber Training: Developing the Next Generation of Cyber Analysts*, 2011.

[75] National Research Council, *Building a Workforce for the Information Economy*, Washington, D.C.: The National Academies Press, 2001; Timothy R. Homan and Zachary Tracer, "ADP Estimates Companies in U.S. Added 42,000 Jobs," *Bloomberg*, August 4, 2010.

cannot judge the quality of their staffs' work, and that they cannot make informed decisions. This undesirable situation is illustrated in Figure 4.1 for a staff member who, after years as a technical staff expert (yellow), has mastered the content of the discipline (blue curve), and then transitions into a management role (orange). As a manager, this person's skills stagnate and do not keep pace with the content of the discipline. It therefore becomes difficult to effectively manage and lead staff.

Figure 4.1. Stagnant Skills Incompatible with Rapidly Evolving Discipline

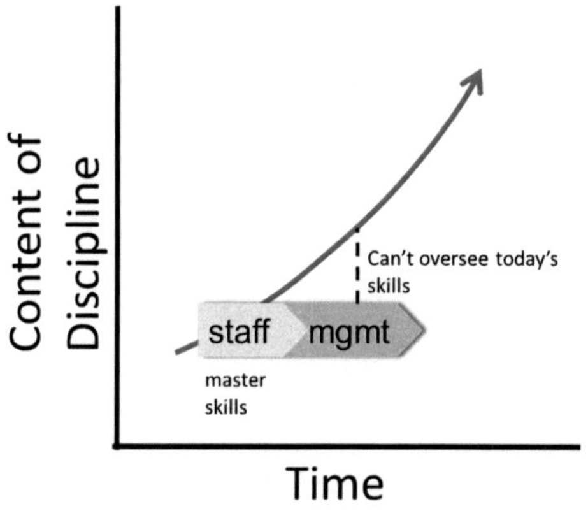

Based on our interviews, companies described a two-step approach to avoiding this undesirable situation. First (the left side of Figure 4.2), they reduce the rate of change of the skills required to master the content of the discipline. Companies reported that they did so by allowing their low- to midlevel managers to specialize in one area and progress in their careers within that area, thus limiting the variation of technical areas they need to manage. Second (the right side of Figure 4.2), companies reported that they maintain the technical depth of their managers, allowing them to continue to keep up with the pace of change of the discipline and make good management decisions.

Again, according to our interviews, by taking these steps, managers reported that they can step into jobs their staffs do, if necessary, although not with the same efficiency. Managers described the need to maintain this technical competence to allow them to identify good work and bad work, and take the appropriate steps as a result. Furthermore, when more-integrative and more-complex decisions need to be made at higher echelons of command, the managers have the technical depth to make those decisions well. This is especially important in IT organizations that retain decisionmaking at higher, centralized levels.

Figure 4.2. Commercial Practices to Promote Technical Depth of Leadership

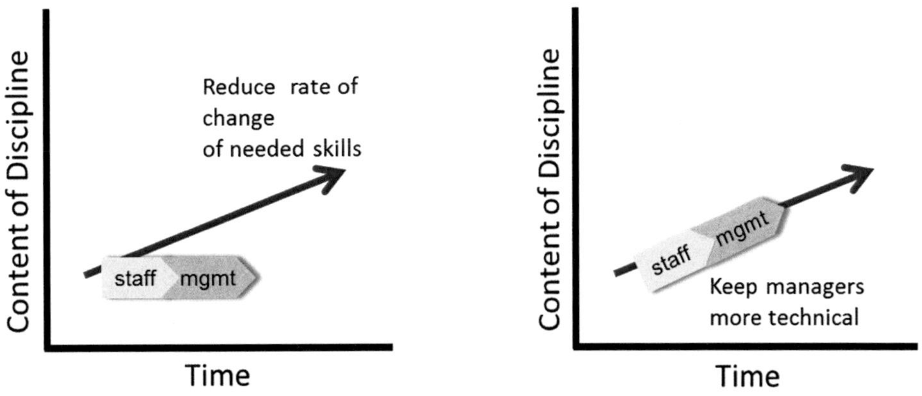

Organizational Strategies Can Encourage Technical Depth

The technical capabilities of staff and leaders can be strengthened by an appropriate organizational strategy. In particular, a consolidated, functionally aligned organization, where personnel with similar tasks and skill sets are grouped into one unit, is most effective at encouraging technical depth. This is because "all human knowledge and skills with respect to specific activities are consolidated, providing a valuable depth of knowledge for the organization. . . . [And this] also promotes in-depth skill development of employees."[76]

On the other hand, a product-oriented organization (consisting of cross-functional teams of specialists) might "eliminate in-depth competence and technical specialization" of those specialists, particularly over time.[77] This regression to the mean is a challenge faced by organizations with few technical experts who are working in isolation from their peers. As such, when organizations seek to maintain a workforce that has technical depth, it is better to consolidate the organizational structure (e.g., at the corporate level), rather than dispersing smaller groups of specialists throughout the different business units.

Figure 4.3 depicts the above discussion graphically for staff with the generic functions F1, F2, and F3. In the consolidated functional organization on the left, there are more personnel adept at each function within the same organization, and formal lateral ties connect similarly skilled staff serving liaison roles in other units to the center of gravity of this function at the headquarters unit. This arrangement encourages technical depth of staff and leadership. If, instead, personnel adept at each function were divvied up among many units (e.g., product-

---

[76] Daft, 2008.

[77] Daft, 2008.

oriented units), as depicted on the right, the small numbers of experts in each unit and the lack of ties to their peers would not encourage development of technical depth in the same way.

Figure 4.3. Organizational Strategies Can Influence Technical Depth

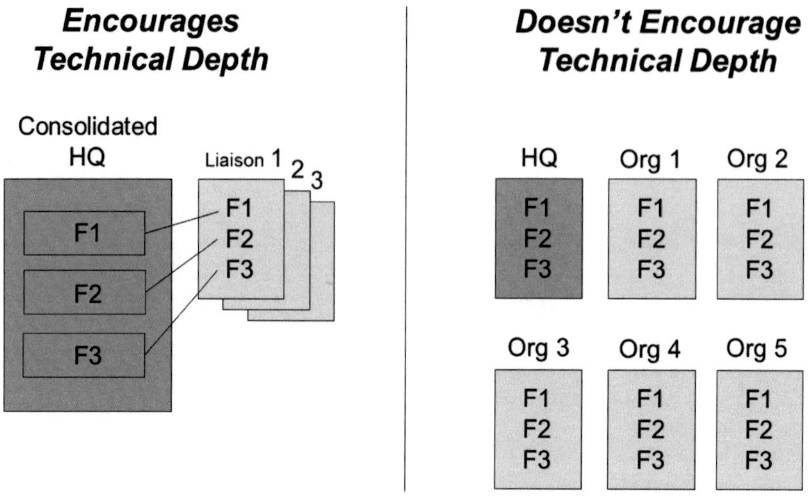

Commercial Model Leverages Many Interconnected Practices

Based on our interviews, we find that the key commercial practices we have described to this point are interconnected and together form an overall commercial model for cyber organization and workforce management. Figure 4.4 depicts all the components of the model, which is displayed as a circle to emphasize that it is both iterative and dynamic, with companies repeatedly tracing this path as technologies and practices evolve.

As technologies evolve, a field of practice emerges (the top of Figure 4.4). For example, new technologies are introduced—e.g., the personal computer, the BlackBerry—and those technologies are adopted by forward-looking business units to improve their productivity. Or other innovations come along—e.g., cyberattacks stealing intellectual property—that cause the corporation to react in other ways. As a result, specializations (e.g., system administrator, malware analyst) emerge and are refined over time to address these new business needs. As more business units adopt similar technologies or share common needs, networks and systems are standardized across business units to gain efficiencies, and this allows for the consolidation of processes. Ultimately, consolidated management structures emerge, aligning like specialties under single organizations (e.g., IT and InfoSec). These large organizations then foster technical depth in their personnel and allow for the growth of leaders with technical depth—leaders who are able to make complex decisions with clarity. Finally, with technical leadership and staff in place, the decision can be made to outsource any noncore capabilities that present opportunities for economic benefit, with the confidence that strong technical oversight will be in place.

Figure 4.4. Interconnected Practices Form a Commercial Model for Cyber Organization and Workforce Management

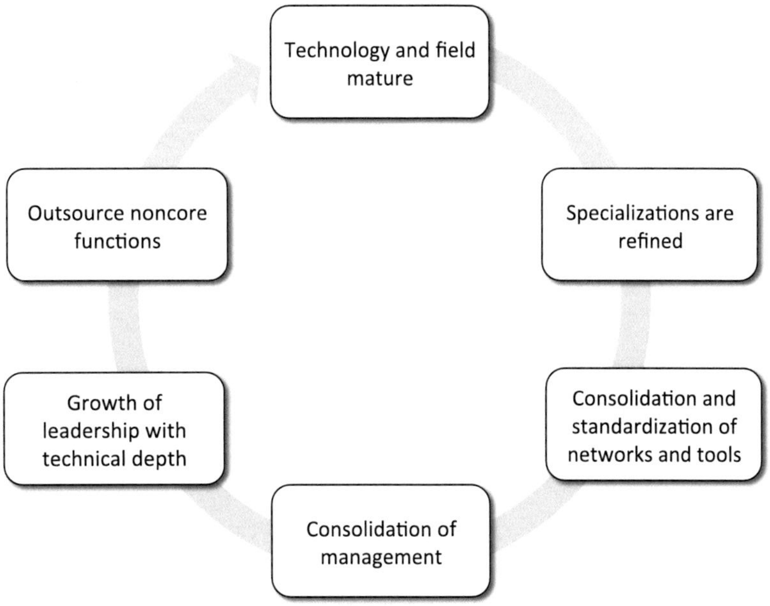

After a field and its technology are centralized and standardized, the next wave of technology innovation can occur, which does not fit into any of the established organizations or standards. This then leads to small niche areas within an organization that adopt the new technology and apply it in a specialized way with no precedence. This restarts the cycle, as the new technology eventually gets adopted by a wider group of people, and gradually what used to be the new technology becomes well established and its use becomes standardized and centrally managed. This cycle continues again as new innovations or other changes in the operating environment occur.

Because of the differences in the level of their maturity, IT and InfoSec are on different iterations of this spiral process. In fact, several themes we observed were effectively pushing InfoSec to a more mature state, like IT. For example, in many of our interviews, companies described a desire to automate more and more of the InfoSec process. Additionally, there was a push toward creating a "cyber playbook" that was akin to the checklists used by help desks.

This means that as long as there continues to be innovation in IT and InfoSec, there will be no one best, stable way to manage them. The way they should be organized will depend on how mature the field and technology are, so companies should always be vigilant and periodically reassess whether their approaches are consistent with where the field and technology stand.

5. Traditional Practices Predominate for Recruiting and Retention

Like USAF, most of the companies we interviewed preferred to hire early-career staff and retain them for decades. Consequently, are the often-voiced government concerns of a cyber-workforce shortage also felt in the commercial sector?[78] And if so, have companies had success with nontraditional approaches, such as cyber competitions and retention incentives, or do they employ other practices?

We describe commercial practices associated with recruiting and retention for IT and InfoSec personnel based on interview results and the literature.

Companies Recruit Recent STEM Graduates from Good Colleges

Most of the large companies we interviewed described the importance of hiring staff with a bachelor's degree. First, such a degree can indicate an applicant's ability to persist, and ultimately succeed, despite challenges. Second, it indicates that the applicant is likely to possess the right attitudes and behaviors to function in a professional work environment—e.g., the propensity to share information and cooperate with coworkers for the good of the company. Several of the organizations we interviewed highlighted the importance of a candidate's organizational fit and cultural behaviors, alongside technical skill requirements, as part of their selection criteria.

Furthermore, companies tended to actively recruit those with degrees in science, technology, engineering, and mathematics (STEM) fields, especially computer science, InfoSec, computer engineering, and electrical engineering. While companies certainly hire from nearby regional universities, nationally known "good schools" for cyber are also valued. Several interviewees mentioned selecting applicants from colleges designated as an NSA/DHS National Center of Academic Excellence.[79]

The preference for personnel with bachelor's degrees is perhaps surprising, and certainly contrasts with USAF workforce requirements (enlisted personnel—the vast majority of USAF

---

[78] According to other RAND work, there is a "rising difficulty of finding and retaining qualified individuals at what are considered reasonable wages . . . at the high end of the capability scale: roughly the top 1–5 percent of the overall workforce. These are the people capable of detecting the presence of advanced persistent threats, or, conversely, finding the hidden vulnerabilities in software and systems that allow advanced persistent threats to take hold of targeted systems." Martin C. Libicki, Dave Senty, and Julia Pollak, *Hackers Wanted: An Examination of the Cybersecurity Labor Market*, Santa Monica, Calif.: RAND Corporation, RR-430, 2014.

[79] National Security Agency and Central Security Service, 2009.

personnel—are not required to have a college degree).[80] As discussed in Chapter Three, the companies we interviewed tended to outsource tier 1 help desk functions, and, therefore, those personnel are not included in this finding; it is possible that such staff do not predominantly have bachelor's degrees.

Cyber Competitions

Some of the companies we interviewed—particularly those selling services to the U.S. government—described cyber competitions as a way to identify people with competence or skill in InfoSec. Across the board, however, most companies approached cyber competitions with some caution, preferring to let others blaze the path with this recruiting method. They reported that this wait-and-see approach was possible since they did not perceive problems with their current recruiting practices.

On the other hand, a few companies sponsored their own internal competitions to identify existing staff with both skill and interest in InfoSec. By screening existing staff, companies might be able to identify untapped talent who could then be trained as InfoSec professionals.

Competitions are often touted for increasing the pool of applicants into cyber fields by generating interest among middle school and high school students, then gradually growing them into capable job candidates year by year. The academic literature indicates that competitions attract "experienced individuals who will remain in the profession for the long-term," but their ability to attract new entrants to the field is unknown.[81] That is, there is some concern that competitions might attract exactly the same people who would have selected cyber careers on their own, even without the lure of competitions.

Midcareer Professionals with Demonstrated Experience Are Also Valued

According to our interviews, smaller companies that do not have the resources to internally train employees to develop needed specialized skills seek out midcareer professionals, with ten or more years of relevant work experience. For these organizations, demonstrated experience is the most important consideration in hiring decisions. We heard repeatedly that offensive and defensive military cyber experience was highly sought after because companies believed that these individuals had received good training in the military. Some companies described the pool of midcareer candidates as smaller than the pool of recent college graduates.

---

[80] While enlisted personnel are not required to hold college degrees, 63 percent have completed some college coursework, and nearly 8 percent hold bachelor's degrees, according to Air Force Personnel Center data; see Air Force Personnel Center, 2014.

[81] David H. Tobey, and Portia Pusey, and Diana L. Burley, "Engaging Learners in Cybersecurity Careers: Lessons from the Launch of the National Cyber League," *ACM Inroads* 5, no. 1 (March 2014).

The cutting-edge cybersecurity companies we interviewed emphasized the importance of hiring people who are "known quantities," as demonstrated by their participation in open source communities, online forums, or ethical hacker groups. Furthermore, as relatively small companies, they can rely on leveraging the personal and professional networks of their existing employees to identify promising candidates.

Pay Is Not the Sole Driver of Retention

Retention of highly skilled employees is important for companies that want to maintain their competitive edge. In particular, there are high costs for having to hire new people and groom them to be the high performers who drive business success.[82] One report in 2007 found that the average Fortune 500 company improved earnings by almost 15 percent by improving talent management.[83]

Although the amount of pay can be an important aspect of retaining top talent, there is academic literature suggesting that it should not be considered the sole factor. In particular, advancement opportunities, good colleagues, job satisfaction, the way the organization treats its employees (i.e., with respect), and organizational prestige all lead high-performing employees to stay.[84]

In fact, the median salary of IT and InfoSec professionals are in line with the pay and benefits received by military personnel. To accurately compare military pay with civilian pay, we used the Regular Military Compensation (RMC) Calculator.[85] The RMC Calculator includes basic pay, basic allowance for subsistence, and basic allowance for housing. Additionally, since most allowances are tax exempt, the RMC Calculator includes those built-in tax advantages.[86] The median cyber officer in USAF is an O-3 with approximately 10.6 years in service. This grade and time in service has a civilian equivalent salary of $94,995.04. For enlisted personnel, the median is E-5 with 8.7 years of service, which results in a civilian equivalent salary of $55,948.77. For occupations that required a bachelor's degree, we compared the median civilian salary with the median officer salary, although we acknowledge that enlisted personnel are often

---

[82] Oracle, *Talent Retention: Six Technology-Enabled Best Practices*, 2012; ADP, *Effective Talent Management Has Become an Essential Strategy for Organizational Success*, Alpharetta, Ga.: ADP, Inc., 2010.

[83] Hackett Group, "Hackett: Companies Can Improve Earnings Nearly 15% by Improving Talent Management Function," July 24, 2007.

[84] John P. Hausknecht, Julianne Rodda, and Michael J. Howard, "Targeted Employee Retention: Performance-Based and Job-Related Differences in Reported Reasons for Staying," *Human Resource Management* 48, no. 2 (2009).

[85] Office of the Under Secretary of Defense for Personnel and Readiness, "Regular Military Compensation Calculator," undated.

[86] We assumed a family size of one and a state marginal tax rate of 0 percent, which produces the most conservative estimate.

fulfilling those roles within USAF. Table 5.1 details the median civilian salary for various IT and InfoSec job roles and the equivalent median military compensation. The data indicate that current military officer compensation exceeds the private sector.

Table 5.1. Civilian and Military Compensation Comparison for IT and InfoSec Jobs

Career Field	Bureau of Labor Statistics Occupation	2012 Median Civilian Pay ($)	Regular Military Compensation ($)
InfoSec	InfoSec analysts	86,170	94,995
IT	Computer support specialists	48,900	55,949
IT	Network and computer system administrators	72,560	94,995
IT	Database administrators	77,080	94,995
IT	Computer network architects	91,000	94,995

SOURCES: Bureau of Labor Statistics, 2014c; Office of the Under Secretary of Defense for Personnel and Readiness, undated.
NOTE: The table underestimates full military compensation since it does not take into account health care or retirement advantages over the civilian workforce.

There are many things that companies can do to retain and train top talent. Some examples include: (1) rotate high performers into various positions throughout the company to maintain interest; (2) include nontechnical training (e.g., focusing on the business); (3) provide opportunity to interact with top management; (4) support employee's passion for technology; and (5) facilitate exposure outside of the company.[87] Another strategy is to develop a mentoring program as a way to save money, retain workers, build leadership, and grow talent.[88]

Our interviews indicated that companies did indeed take many of these approaches to retain their top talent. Several interviewees reported that people stay because the work is interesting, rewarding, and cutting-edge. One company sent its people to conferences to keep up-to-date with the latest technologies and trends. Some companies also rotated their employees through different business units to get exposure to different areas of the company and keep employees interested. As a result, most companies we interviewed reported good retention rates (less than 10 percent attrition per year). However, other RAND research indicates that personnel with "elite" InfoSec skills might be in short supply, and, therefore, there is upward pressure on salaries to retain staff with this skill set.[89]

---

[87] James Kaplan, Naufal Khan, and Roger Roberts, "Winning the Battle for Technology Talent," McKinsey & Company, May 2012.

[88] Claire Schooley, Connie Moore, and Ralph Vitti, *Drive Employee Talent Development Through Business Mentoring Programs*, Cambridge, Mass.: Forrester Research Inc., 2010.

[89] Libicki, Senty, and Pollak, 2014.

Nevertheless, we noticed a difference between the public and private organizations we interviewed. In particular, some public organizations mentioned that they are unable to offer the kinds of incentives the private sector can. But other public organizations mentioned that security and stability of a government job helped with retention. Therefore, public and private organizations might have different advantages and disadvantages in offering incentives for recruitment.

6. Commercial Practices Might Aid USAF

The preceding chapters described commercial practices found in companies that we selected for study based on their similarity to USAF; these practices were further supported by the literature. Several USAF practices are already similar to commercial practice—namely, standardization of IT processes; a large in-house IT workforce performing a critical core function; consolidated management of InfoSec; small, empowered InfoSec teams; and a move toward preferring officers with STEM degrees. We will not further belabor those topics. Instead, we consider whether the remaining commercial practices are applicable to USAF.

When assessing which commercial practices are likely to be applicable to USAF, we took the following approach. We contrasted the commercial practice with the constraints USAF faces. For example, commercial practices for workforce management leverage the ability to hire midcareer professionals, but USAF is constrained from completely implementing such a practice given that military personnel are most often accessed as junior service members, straight out of high school or college. Constraints like these are, in most cases, not easy for USAF to remove or relax. Therefore, we identify how the commercial practice would likely change if subject to the same constraints as USAF.

USAF Has Unique Constraints Not Experienced in the Commercial Sector

It is important to recognize that several of the conditions in which USAF operates differ from those found in the commercial sector, and might, therefore, recommend against the application of commercial practices. The constraints we identified as potentially troublesome are listed across the top of Table 6.1. First, within USAF there is staff turnover due to the frequent permanent change of station (PCS) cycle. This is in contrast to the relative stability seen in the commercial sector. Second, a significant number of people in the military are not required to, nor do they, possess bachelor's degrees. Generally, commercial sector firms prefer college-degreed staff as a way to ensure a professional workforce and, furthermore, to select staff with demonstrated abilities in relevant academic disciplines. Third, the "up or out" nature of career progression in the active component places unique pressures on developing airmen who are "promotable" when compared with their peers in other career fields. This dynamic is sometimes said to reward generalists over technical specialists.

Table 6.1. USAF Constraints and the Commercial Practices They Might Affect

	USAF Constraints						
Commercial Practices	PCS Cycle	No Degree	Up or Out	Contested and Deployed	Customer Boundaries	Accession Model	Offensive Ops
IT and InfoSec job managed separately			X				
InfoSec decisionmaking decentralized	X	X					X
Consolidated IT and InfoSec				X	X		
95% IT, 5% InfoSec				X			X
Limited outsourcing				X			
Technical leadership			X				
STEM degrees and midcareer hires		X				X	

Fourth, USAF is expected to operate in contested *and* deployed environments with little warning and preparation. While several of the organizations we interviewed described operations in contested *or* deployed environments, few are required to tailor their operations to excel in the type of doubly restrictive environment for which USAF is optimized. In fact, it is arguably most important that USAF function effectively in deployed environments while under cyberattack (and physical attack) to fulfill its warfighting missions—this is at the heart of the need for USAF's existence.

Related to this, the fifth constraint is the way in which USAF presents forces to "customers"—e.g., regional combatant commands (RCCs) and major commands (MAJCOMs)—and thus must navigate the complex boundaries between them. For example, one RCC might be at war and under cyberattack, while others are engaged in peacetime operations. This necessitates the ability to provide tailored services in each region.

Sixth, although the USAF civilian corps and the reserve component have some capability to hire midcareer professionals, the active component typically does not access midcareer experts from outside USAF. This policy limits USAF access to experienced technical talent, whereas commercial practices value infusion of midcareer talent when companies need to quickly "buy" skills that are not resident within the organization. However, USAF does have the latitude to exercise force shaping and define manpower requirements and roles in both the active and reserve components. In this respect, the defensive, mission assurance orientation of InfoSec (with an experienced and stable cadre of talent) might be appropriate for a focused responsibility of the reserve component, and civilian subject-matter experts, in supporting the cyber mission.[90]

Finally, USAF's defensive cadre must work in concert with its offensive counterparts. For example, the defensive personnel must be prepared for a counterattack as the result of an offensive operation. Since the commercial sector is legally prevented from engaging in offensive cyber operations, its cyber workforce management is less informative regarding the holistic approach that USAF might take to both offensive and defensive cyber operations.

If Subject to USAF-Like Constraints, Commercial Practices Would Likely Change Only Marginally

Table 6.1 also lists, down the left column, the commercial practices we found to be potentially affected if confronted by USAF-like constraints. In what follows, we describe the nature of the interaction between commercial practices and these constraints, as well as the way in which the commercial practice would likely change if subjected to the constraints.

IT and InfoSec Are Managed as Distinct Disciplines

The first three practices—maintaining different job roles, training regimens, and career paths for the two distinct fields of IT and InfoSec—are fundamental to commercial workforce management and organizational strategies. However, a constraint frequently claimed to be in conflict with such practices is the need to promote junior officers at a sufficient rate to retain them, given the up-or-out pressures they face. That is, as indicated in Table 6.1, it might seem to be difficult to sustain large numbers of leaders in a small career field like InfoSec, yet the up-or-out pressures necessitate finding a way to provide sufficient numbers of leadership positions to house the rising leaders or risk losing them.

Recall, though, that commercial InfoSec organizational approaches emphasize the need to maintain few staff per supervisor in many small, high-performing teams. The prevalence of such teams presents the opportunity for leadership positions that are no less demanding than leading larger teams of IT staff. Therefore, maintaining many leadership positions does not appear to be in direct conflict with the practice of maintaining InfoSec as a smaller specialty distinct from IT.

Another aspect of this constraint is the desire to produce a certain number of very highly ranking leaders (e.g., general officers). Here again, we do not see commercial practices as necessarily incompatible with this desire. Even in the commercial sector, very senior leaders (e.g., CEOs) are not usually narrow experts who came up from a single discipline alone—at some point in the careers of these few leaders, the value of business acumen, innovation, and management skill trumped that of subject-matter expertise. The need for identifying and

[90] Libicki, Senty, and Pollak, 2014.

developing these exceptional, differently skilled individuals does not invalidate the necessity for maintaining two different cadres of skilled workers; it simply argues for efficient approaches to identify the few such differently skilled individuals and provide them the opportunities required to grow.

Decentralized Decisionmaking for InfoSec

Decentralizing decisionmaking for InfoSec allows high-performing, cross-functional teams of experts to apply their knowledge to react quickly to address rapidly evolving security needs. In the commercial sector, these experts develop the technical depth necessary to be entrusted with decisionmaking over the course of many years of on-the-job training; they master the nuances of the company's systems and come to understand the threats that hold them at risk. Longevity is particularly important for elite InfoSec teams, such as CSIRTs.

As indicated in the second row of Table 6.1, the constraint of a rapid PCS cycle could interfere with the development of such expertise unless the jobs that personnel rotate among are similar enough to allow significant transfer of knowledge from one job to the next. Furthermore, entrusting recently enlisted high-school graduates with decisionmaking would not be appropriate if they have yet to demonstrate the critical thinking skills necessary to make those decisions. Therefore, decentralized decisionmaking for InfoSec should be practiced only if the experience of the personnel and the effect of the PCS cycle are taken into consideration—for example, by requiring that the jobs with the deepest specialized skills be filled with seasoned personnel who do not rotate as frequently.

Additionally, in an environment where offensive and defensive operations might need to work in concert to achieve a desired objective, decentralized decisionmaking might inadvertently undermine the mission unless there is proper coordination across teams.

Consolidated Organizations

The commercial practice of consolidating IT staff under a single organization headed by the CIO leverages standardization across the company to gain efficiencies and promote technical depth. Likewise, consolidating the management of InfoSec within a single organization headed by the CISO reaps benefits, including more-effective network defense within a well-monitored boundary. However, the constraint (indicated in Table 6.1) of needing to provide tailored IT and InfoSec services to each RCC or MAJCOM calls into question the feasibility of consolidating IT and InfoSec too broadly. However, we observed similar constraints in the commercial sector—i.e., the need for IT and InfoSec to be mindful of the boundaries separating subsidiaries of large, multinational conglomerates, stemming from the need to comply with different regulatory regimes. Therefore, the constraint of needing to provide tailored support to different "customers" does not appear to alter the commercial practice of consolidation—it merely limits the extent of consolidation (e.g., to the subsidiary level).

Furthermore, consolidation of IT in the commercial sector was attributed with yielding efficiencies through economies of scale, thereby reducing the size of the IT force required to support the company as a whole. Based on organizations with a similar complexity as USAF, we observed a ratio of one IT staff member for 25 employees. If this commercial practice was applied to a corporation the size of USAF, we would expect an IT cadre of approximately 14,000. Of course, this is far smaller than the size of the USAF IT cadre.[91] And it is *not* likely that USAF could halve its IT force by gaining efficiencies, because of the limitations on consolidation. In fact, commercial organizations that must respect regulatory boundaries end up retaining more staff than those that can consolidate to the maximal extent.

Furthermore, if also faced with the constraint that the organization must persevere in deployed and contested environments, practices would need to adapt so as to allow for increased resilience at the deployed locations. Both of these constraints reduce the extent of efficiencies that can be generated by consolidation. Therefore, while we expect that commercial practices that consolidate to the greatest extent practical (without harming effectiveness) would aid USAF efficiencies, the extent of efficiencies reaped in the commercial sector will *not* be available to USAF given these constraints.

95 Percent IT, 5 Percent InfoSec

We next scrutinize the relative proportions of IT and InfoSec personnel that were so consistent in commercial practices (95 percent IT and 5 percent InfoSec). It is unclear if the 95 percent and 5 percent practice will stand in the face of USAF constraints, such as those shown in the fourth row of Table 6.1. Given USAF's need to deliver effective IT in contested, deployed environments, these operating conditions place an imperative on both IT and InfoSec capabilities. However, the commercial ratio might still be informative because, given this imperative, USAF should be postured to withstand threats at least as robustly as the commercial sector. Additionally, given that USAF's offensive mission might result in additional demands on its DCO forces, which are not present with commercial companies, USAF might require an InfoSec force greater in size proportionally than seen in commercial entities. Consequently, commercial practice should be considered a *lower bound* for USAF force structure planning assessments.

A quick comparison indicates that the proportions of IT to InfoSec in USAF are light on InfoSec, when compared with the commercial practice. Given a commercial organization with 36,000 IT personnel (the approximate size of USAF's IT cadre), implementation of commercial practices would predict an InfoSec cadre of 1,895 personnel. This is 2.3 times larger than the

---

[91] These calculations are based on a 350,000-member organization. The size of the USAF IT cadre is approximately 36,000, according to data provided by AF/A3C/A6C (current as of April 2014).

USAF InfoSec cadre.[92] Alternatively, if we apply these ratios to an organization with 834 InfoSec personnel (the approximate size of USAF's defensive-focused InfoSec cadre), implementation of commercial practices would predict an IT cadre of 15,846 personnel.[93] Again, this is 2.3 times fewer than the current USAF IT cadre.[94] While we do *not* expect that USAF should decrease its IT cadre by a factor of two, the relative level of effort between IT and InfoSec for USAF, as compared with commercial practices, is worth evaluation by USAF. The implications of these differing characteristics would seem to reinforce the need for a cyber manpower review to determine if the InfoSec workforce is adequately sized and if the recent increases as part of U.S. Cyber Command initiatives are bringing the cyber corps into balance.

Limited Outsourcing

Recall that commercial practice is to limit outsourcing to functions that are deemed to not be core capabilities, and to also require robust in-house technical capability to oversee the contracts. If faced with the additional constraint to excel in heavily threatened, deployed environments, we expect that commercial practice would need to be altered to become even more cautious about outsourcing. Functions supporting deployed forces might be declared critical core capabilities, and thus removed from the possibility of outsourcing. However, we expect that outsourcing conducted in support of home station activities and to provide nascent capabilities out of scope of in-house personnel would remain similar to existing commercial practices.

Technical Leadership

Valuing and cultivating leaders with technical depth is a hallmark of commercial practice. Particularly at the lower levels, IT and InfoSec leaders appear to be evaluated without much comparison with their peers in other fields and parts of the company. However, if presented with the constraint that commercial cyber staff must be promotable at similar rates as their peers in other fields, it is not clear if the practice of valuing technical depth in cyber would change or not, particularly since it is difficult to envision a reason for such a constraint emerging in the commercial sector. What is clear is the value that technical leaders provide to the company and the practice of providing rising superstars and senior managers with opportunities to broaden their company knowledge by applying their skills in other business units further supports this emphasis on providing value to the company.

---

[92] To draw a direct comparison with the commercial sector, we limited the USAF InfoSec cadre to those conducting defensive operations, not offensive.

[93] This is according to data provided by AF/A3C/A6C (current as of April 2014).

[94] InfoSec personnel counts exclude USAF OCO and cyber intelligence, surveillance, and reconnaissance staff. Data were provided by AF/A3C/A6C (current as of April 2014).

However, the practice of maintaining a technical track for staff to pursue as an alternative to a career in management would clearly need to be revisited if career progression and salary increases were constrained to be solely tied to progression into management roles. Fortunately, this is only a constraint on USAF officers, as the enlisted force can effectively pursue an "up and stay" approach, which is more consistent with the commercial practice of maintaining a technical track.[95]

STEM Degrees and Midcareer Hires

Commercial practice relies on the university system to provide the fundamental education and professionalism that personnel will need to effectively participate in the workforce. This practice would require significant modification if presented with the constraint that most hires will not have bachelor's degrees (see the last row in Table 6.1). These modifications would likely take at least two tacks. First, selection criteria would need to be substantially revised to develop a means to identify candidates who will be able to cope with the rigors of the job and work within the norms of the corporate culture. Specific to cyber, such selection criteria might include cyber aptitude testing or might look for signs of an interest in and affinity for cyber roles, such as participation in cyber competitions during secondary school. In fact, commercial practice for elite cybersecurity jobs employed such tests of aptitude and ability *in addition* to formal educational requirements. This practice would likely become more important when employers are unable to rely on educational credentials to vouch for the skill of applicants. Second, increased training regimens would be needed to instill the subject-matter knowledge typically acquired during an undergraduate degree program. This training would be in conjunction with existing specialization training common in commercial practice today. Furthermore, given the high washout rates of college students initially selecting STEM fields,[96] USAF might also need to assess aptitude and fortitude in the types of disciplines most relevant to cyber (e.g., computer science, information systems, and computer engineering).

In addition to hiring recent STEM graduates from good colleges, commercial practice includes the ability to hire midcareer professionals when in-house capabilities do not exist in sufficient numbers. This is particularly useful when the organization is entering a new environment that requires cyber skills not needed before; instead of waiting to develop these skills in-house, companies might hire existing experts. If presented with the constraint that this

---

[95] However, compared with personnel in technical tracks in the commercial sector, there are more pressures placed on the USAF enlisted force to progress to higher ranks with management responsibilities, as governed by the "high year of tenure" policies.

[96] "A total of 48 percent of bachelor's degree students and 69 percent of associate's degree students who entered STEM [degree programs] between 2003 and 2009 had left those fields by spring 2009," according to a study conducted by the U.S. Department of Education. Xianglei Chen and Matthew Soldner, *STEM Attrition: College Students' Paths into and out of STEM Fields: Statistical Analysis Report*, Washington, D.C.: U.S. Department of Education, November 2013.

practice could no longer be supported to the same extent, commercial practice could adapt in at least two different ways. One way would be to leverage external training programs to increase the skills of staff currently inexperienced in this new field. Another would be to rely more on outsourcing to companies who do possess these skills. However, given the cautious approach to outsourcing, it is likely that both routes would be pursued in combination to ensure that the transfer of knowledge from the contract provider to the in-house staff is likely to succeed.

Other Commercial Practices Are Unaltered by Constraints

Note that several commercial practices are not included in Table 6.1 because we found no adverse interactions with USAF-like constraints. Those practices—which should be directly applicable to USAF—are:

- assigning many staff members per supervisor for IT and few staff members per supervisor for InfoSec
- use of standardized processes for IT
- preference for creative thinkers for InfoSec missions
- establishment of strong lateral linkages between IT and InfoSec organizations
- assigning IT staff with similar functions to the same organization (functional alignment)
- assigning InfoSec staff with cross-functional skills to mission-focused organizations (divisional alignment)
- retaining IT as a critical core competence with a large, in-house staff
- structuring organizations to cultivate staff and leaders with technical depth
- retaining staff with mechanisms other than just pay increases.

In the final chapter, we discuss how USAF might go about adopting the (adjusted) commercial practices we have described.

7. Options for USAF to Implement Commercial Practices

We found strong parallels in the commercial sector for USAF DoDIN Ops and DCO activities, and—although none of the companies we interviewed were as large as USAF or were required to excel in deployed, contested operating environments—we find that the commercial practices we identified are likely to be adaptable to USAF and deliver a measure of effectiveness and efficiency gain that USAF would find beneficial.

In this report, we described the commercial practices we identified for cyber workforce management that are supported by both theory and practice. In the previous chapter, we described how they are largely applicable to USAF already or could be adapted. Next we offer options for USAF to adopt these commercial practices to help improve workforce management. We recognize that there might be other considerations that would preclude USAF from adopting all these options; however, we offer them as specific issues to investigate.

Align Career Fields with Either IT or InfoSec

Acknowledging that these are different disciplines that require different management approaches is a key contrast between commercial practice and current USAF practice. Without the ability to manage these disciplines individually, USAF will not be able to take advantage of many of the other commercial practices that hinge on this foundational element.[97] Therefore, USAF should evaluate whether AFSCs for officers and enlisted personnel, and career designators for civilians, could be aligned to focus on a single specialty—IT *or* InfoSec. This would require an assessment of the roles and responsibilities, as well as entry qualifications and training plans. Customizing training for each specialty to develop and retain technical depth and currency should aid both the efficiency and effectiveness of cyber professionals. Furthermore, staff assignments could be made within staff members' specialties to retain depth and currency; breadth could be gained by using staff members' specialties at assignments throughout the USAF or joint community. In particular, while a cursory evaluation indicates that many USAF enlisted AFSCs are aligned to one specialty, other enlisted and officer specialties appear to require an individual to master both fields. Given the rapid pace of technological change and the complexity of many of these activities, USAF should analyze the benefits of aligning AFSCs to either IT or InfoSec.

Furthermore, USAF should evaluate the extent to which those IT and InfoSec roles that require the greatest technical depth and longevity could be filled by civilians, guard, and reserve personnel. Associating highly technical InfoSec jobs in a way that either ensures longevity in

[97] Recall from Figure 4.2 that maintaining specializations is foundational to many other commercial practices.

these positions or rotates personnel among very similar positions might deliver both efficiency and effectiveness gains for USAF. Instituting recurring training (and/or certification) opportunities to ensure that staff are given the opportunity to remain current as technologies change should also be explored.

Increase USAF InfoSec Workforce

The size of the USAF InfoSec workforce is 2.3 times smaller than one might expect based on commercial practices. It seems clear that an organization such as USAF should be postured to withstand threats at least as robustly as the commercial sector. We would also expect a larger USAF InfoSec workforce due to the increased demands of operating in a deployed and cyber-contested environment. Further increases in size requirements might come from USAF's offensive mission, which might result in additional demands on its DCO forces not present with commercial companies.[98] Consequently, commercial practice should be considered a *lower bound* for USAF force structure planning assessments.

The implications of these differing characteristics would seem to reinforce the need for a cyber manpower review to determine if the InfoSec workforce is adequately sized to meet the USAF cyber mission.

Retain IT as an Essential Core Capability

Projections suggest that 94 percent of the commercial cyber workforce will be in IT through 2022, reflecting the view that IT remains a core capability critical for success in the commercial sector. As USAF is also heavily reliant on command, control, communications, and computer capabilities, we expect that the need for IT capability to persist for USAF as well. Commercial practice limits the outsourcing of many core cyber functions to protect the company's ability to execute its mission. USAF should similarly identify which IT and InfoSec functions are critical to USAF missions and carefully assess the risk of outsourcing those core capabilities. The commercial ratio of 94 percent of the cyber workforce engaged in IT could serve as an interim planning factor until this USAF assessment is completed. Particularly with the advent of the Joint Information Environment (JIE), USAF should analyze the demand for IT professionals within its ranks. This assessment should also evaluate the extent to which IT roles are inherently military, or whether they could be filled by government civilians.

---

[98] Additionally, if USAF is to count other cyber functions (e.g., OCO and cyber intelligence, surveillance, and reconnaissance) among its InfoSec workforce, then the percentage could be expected to be larger still.

Access Cyber-Capable Personnel

Identifying individuals with existing cyber skills or the aptitude to quickly master cyber skills is a hallmark of commercial practice. And while the commercial sector leverages the university system to identify promising applicants, USAF must perform more of the vetting itself. To improve the accession of cyber-capable personnel, USAF should investigate the feasibility of establishing and implementing tests for IT aptitude and InfoSec aptitude as part of the enlisted accession process. Additionally, a part of the officer accession process, USAF should prefer candidates with relevant academic degrees from universities with noted cyber programs (e.g., NSA and DHS cyber centers of excellence) or relevant extracurricular activities (e.g., participation in open source forums, certifications, or experience with cyber contests).

Structure Organizations to Gain Efficiencies and Effectiveness

There are two fundamental ways that USAF could take advantage of commercial practices to improve the effectiveness of cyber organizations and to identify cost savings. First, structure organizations, including cyber squadrons, according to the guidelines laid out in organizational design. As described in Chapter Two, these guidelines apply well to IT and InfoSec and are validated by their use in commercial practice. Instituting such organizational structures could provide the additional benefit of bolstering the technical depth of staff. To adopt these organizational practices, USAF should structure InfoSec organizations with few staff members per supervisor—i.e., many small cross-functional teams aligned by mission, allowing for many leadership positions—and structure IT organizations in large groups by function.

Second, consolidate the management of these organizations to the greatest extent possible to achieve efficiencies. Commercial practice consolidates IT organizations within a corporate (or subsidiary) organization, with liaison units that support the various business units. InfoSec organizations are consolidated within a corporate organization with broad visibility across the corporation. In some cases, complex organizational boundaries require consolidating only to the subsidiary level (instead of a single organization across multiple subsidiaries). In particular, consolidation should be undertaken when systems and processes are common (e.g., enterprise systems). To seek efficiencies while preserving effectiveness, USAF should assess the feasibility of consolidating the leadership of IT operational organizations under one or more (potentially regionally or MAJCOM aligned) organizations to effectively support customers while gaining efficiencies. If such an approach were adopted, the assignment of formal liaisons from the consolidated IT organization to the supported units (e.g., MAJCOMs) would likely be needed to tailor the services provided to the specialized needs of the user communities.

Appendix A. Characteristics of Companies and Organizations Interviewed

We interviewed a total of 26 companies and organizations from a wide range of industry sectors that shared some commonality with USAF, including financial institutions, major manufacturing firms, defense industrial base firms, energy companies, network security specialists, and large government agencies. To preserve anonymity, we are able to provide some company and organization characteristics but not all.

Of the 26 interviews, only 22 directly contributed to our analysis. Those organizations that were not included had limited applicability to USAF, as they predominately specialized in forensics. Of the remaining companies, 15 were for-profit commercial companies, four were nonprofit commercial companies or public-private partnerships, and the remaining three were government organizations. To ensure that we captured worldwide trends, of the 22 companies that contributed to our analysis, only 16 were based in the United States. Of the 15 for-profit companies, about half had fewer than 10,000 employees and the other half had more than 10,000 employees. Table A.1 details how the 15 for-profit commercial companies break out by sector. We also provide additional details on the operating environments of the 22 companies that contributed to our analysis in Table A.2.

Table A.1. For-Profit Commercial Companies by Sector

Sector	Number of Companies
Technology/cyber security	5
Defense sector	4
Financial	2
Manufacturing	1
Telecommunications	1
Oil and gas	1
Conglomerate	1

Table A.2. Operating Environment

	Home Station	Deployed
Permissive	22	3
Contested	22	—

NOTE: Companies and organizations can be in multiple categories.

Appendix B. Semistructured Interview Questions

This appendix provides an overview of questions asked during our interviews. Depending on the company or organization, not all questions were applicable, but the four high-level topics were always addressed to some degree. Generally, each interview was conducted by at least three analysts, each of whom asked questions and took notes. After the interview was complete, a single authoritative document summarized the contents of the interview using all the notes available. The authoritative document from interviews conducted later in the process were also sent to the interviewee for review.

Subsequently, we compared the interview notes to determine which themes were observed in all of the companies. The findings listed in the report were applicable to at least 90 percent of the for-profit commercial companies, with additional support from at least one nonprofit commercial company. Interview notes from government organizations were used as a reference to compare differences with the private sector.

Organizational Questions

- What is the size of your workforce?
- What is the size of your IT workforce (operate and maintain)?
- What is the size of your InfoSec workforce (protect and defend)?
- What does your organization chart look like?
- Who is (in-house or outsourced) responsible for:[99]
  - the design of your network?
  - IT support (e.g. resetting passwords, hardware installation)?
  - network defense?
  - incident response?
  - analysis of threats?
  - conducting forensics on attacks?
  - active defenses (e.g. denial and deception operations)?
  - setting policy?
  - legal advice?
- What is the relative level of effort between the various network-related tasks?
- How do you determine which work to outsource? How do you select a company to conduct it?

[99] If done in-house, where are they in the organization chart and how many people do that job?

Company Strengths

- Using the NICE Framework lexicon, which particular cyber mission area does your company engage in?
  - Operate and maintain?
  - Protect and defend?
  - Collect and denial/deception operations?
  - Investigate?
  - Analyze?
  - Oversight and development?
- How does your company hire and retain skilled staff?
- How does your company foster innovation and agility?
- Does your company invest in research and development?
- Are there other cyber-related practices that contribute to your company's success?

Enablers

- What is the composition of the company?
  - What is the employee to supervisor ratio?
  - How do the different departments interact?
  - Which staff members are specialists, generalists?
  - Which staff members are technicians, professionals?
- What are the approaches to hiring and retaining skilled staff?
  - How do you recruit?
  - What are the key factors you look for when hiring people (e.g., work experience, certifications, academic background)?
  - Do you make nonstandard hiring decisions? Who has that authority?
  - Are all your cyber personnel technical?
  - What are your retention rates?
- What are the approaches to training, educating, and certifying staff?
  - Do you train your employees?
  - If yes, what type of training do you provide and how frequently?
- Describe the hierarchy of skill sets and job progression.
  - What level of seniority is required for proactive defensive actions?
  - By what metrics do you assess an employee's capability?
  - How do you retain capable employees?
- Describe the company's innovation and agility.
  - Do you conduct any cyber research and development?
  - How do you foster innovation within your staff?
  - What level of responsiveness/agility is required for your various cyber tasks?

- How do you do incident response, surge operations?

Organization

- Describe the organization.
  - How do you measure the performance of your organization?
  - Have you ever reorganized? If yes, how many times and for what reasons?

Appendix C. Organizational Design

Organizational design offers different approaches, taking into account the external stimuli discussed in Chapter Two. These approaches involve a number of elements, including departmental grouping, the hierarchy of authority, lateral linkages, the standardization of tasks, and the centralization of decisionmaking. We discuss each element and then describe how to combine these elements into organizational design approaches to deal with environmental stimuli.

Departmental Grouping

The fundamental units of an organization are the employees. Only when employees are brought together and organized in a well-defined and purposeful manner can the organization begin to function properly and mobilize its collective talent and creativity toward meaningful goals. Departmental grouping describes how individuals can be brought together into such meaningful departments.[100] There are three major types of departmental groupings: functional, divisional, and matrix organizations.

Functional organizations are aligned along functional lines—e.g., accounting, engineering, research and development, production, and human resources. This means that people with certain skills are organized together with people of like skills. Such a grouping takes advantage of economies of scale, since each functional unit can learn from others within the unit and can focus on developing and maintaining their skills. A disadvantage is that because of the separation of units along functional lines, communication and coordination *between* different functional units can become difficult and lead to inefficiencies.

A functional organization provides efficiency and can leverage economies of scale. This grouping encourages the development of depth of skill by allowing the workers to focus on their tasks and learn from each other.[101] In low-complexity environments, the span of control tends to be wider, because the supervisor can manage many people efficiently. The organization tends to have weaker lateral linkages, because there are only modest needs to quickly share information between work groups. It is also typically more efficient if tasks are standardized; the tasks that need to be accomplished are relatively well understood and unchanging, so a one-time codification of tasks will help people do their jobs more efficiently and effectively. Decisionmaking tends to be centralized at higher levels in these organizations, because they can

[100] Richard L. Daft, *Organization Theory and Design*, 10th ed., Mason, Ohio: South-Western Cengage Learning, 2008, p. 90.

[101] Daft, 2008.

take the time to synthesize information that rises up through the chain of command to make the best decisions at higher levels.

Divisional organizations are aligned along lines of business, where divisions could be products—e.g., an aircraft division and a consumer electronics division. Here, each product-oriented division might contain its own accounting, engineering, research and development, and human resources departments. By focusing an entire group of people on one or a few product lines, the divisional group can more rapidly adapt to changing customer needs, with solutions tailored to that particular product.

While the divisional organization is better than a functional organization at dealing with high environmental complexity and variability, the downside is that there could be duplication of effort across different product lines. For example, each division has its own accounting and sales teams, even though it might be possible for the teams to divide their attention among multiple product lines and therefore be more efficient with their resources. Or the engineering component of one product line might want to share certain equipment owned by another unit. Furthermore, employees with certain skill sets will not be able to develop the same level of in-depth expertise as a functional organization, because people with the same skill set are more dispersed throughout the different units than in a functional organization.[102]

Matrix organizations are aligned as a hybrid of functional and divisional form, where employees report to two bosses, one divisional and one functional. This type of organization, if managed correctly, can be very flexible and versatile at dealing with highly uncertain and complex environments by pulling together resources in a way that a traditional functional or divisional organizational structure would not be able to. The downside is that realizing such an nimble organization requires both formal and informal coordination across a wide range of the organization, which can be resource intensive and time-consuming.

Hierarchy of Authority

In most organizations, there typically exists a hierarchy of supervisory relationships between the people, so that roles, responsibilities, and accountability are clearly defined and delineated for the effective and efficient functioning of the organization.[103] Given such a hierarchical structure, *hierarchy of authority* describes the reporting relationships between and within units of an organization and therefore includes both vertical and horizontal aspects.[104]

---

[102] Daft, 2008.

[103] Elliott Jaques, "In Praise of Hierarchy," *Harvard Business Review* (January–February 1990).

[104] Daft, 2008, p. 17.

Vertical authority is indicated by the number of levels in an organization.[105] In a large organization, there naturally tend to be more levels simply because of the larger number of people, and in a small organization with fewer people, there tend to be fewer levels.[106]

Horizontal authority, also called *span of control*, is indicated by the number of peers within a given unit that share the same supervisor.[107] Note that the academic literature also describes this in terms of the number of employees per supervisor.

Lateral Linkages

Lateral (or horizontal) linkage is defined as "communication and coordination horizontally across organizational departments."[108] Typically, lateral linkages are not explicitly shown on an organizational chart, yet they can be a very important factor in its effective functioning. Lateral linkage allows information to flow in the horizontal direction in an organizational structure and lessens the information load that must flow higher in the organizational hierarchy.[109] Lateral linkage can take many forms, such as liaisons, task forces, and teams, and their usefulness depends on the situation.[110]

A liaison is usually located in one department and has the responsibility to communicate and coordinate with another department. A liaison usually connects only two departments; when lateral linkage requires three or more departments to communicate and work together, a task force becomes appropriate. A task force comprises representatives from multiple departments and exists as long as a problem remains to be solved. Thus, a task force is a temporary arrangement. Teams (or working groups or committees), on the other hand, are permanent entities composed of representatives from multiple departments to address recurring issues.

---

[105] Dan R. Dalton, William D. Todor, Michael J. Spendolini, Gordon J. Fielding, and Lyman W. Porter, "Organization Structure and Performance: A Critical Review," *The Academy of Management Review* 5, no. 1 (January 1980).

[106] Daft, 2008, p. 17.

[107] Gerald D. Bell, "Determinants of Span of Control," *American Journal of Sociology* 73, no. 1 (1967); Peter M. Blau, "The Hierarchy of Authority in Organizations," *American Journal of Sociology* 73, no. 4 (1968); and John Bohte and Kenneth J. Meier, "Structure and Performance of Public Organizations: Task Difficulty and Span of Control," *Public Organization Review* 1, no. 3 (2001).

[108] Daft, 2008, pp. 95–101.

[109] Jay Galbraith, "Matrix Organization Designs: How to Combine Functional and Project Forms," *Business Horizons* 14, no. 1 (1971).

[110] Galbraith, 1973; Annick Willem and Marc Buelens, "Knowledge Sharing in Public Sector Organizations: The Effect of Organizational Characteristics on Interdepartmental Knowledge Sharing," *Journal of Public Administration Research and Theory* 17, no. 4 (2007).

Standardization of Tasks

The standardization (sometimes also called *formalization*) of tasks is defined as the degree of codification and the prescription of expected behavior of employees in an organization in performing certain tasks.[111] The tasks that employees must perform can vary widely in the degree of codification required.

For example, the work on an automobile assembly line must be done in a systematic and clearly articulated sequence of steps to keep the assembly line going and to produce the targeted number of cars in a given amount of time. It would be a waste of resources if each line worker had to figure out what to do on his or her own. On the other hand, a custom research and development company must be able to tailor its work to the needs and requirements of each individual customer. Standardization will be of little help in this case, because each customer's needs could be very different. In the former example, the tasks must be highly standardized, whereas in the latter example the tasks will be less standardized.

Centralization of Decisionmaking

The centralization of decisionmaking within an organization is the degree to which decisionmaking authority is distributed throughout the organization.[112] One extreme is an organization where only one person at the top, the CEO, makes decisions—decisionmaking rights are concentrated in one person. The other extreme is an organization where all employees have the same decisionmaking rights. These extreme cases would be appropriate only in very special circumstances, if ever, and, typically, decisionmaking is distributed somewhere in between those two extremes.[113]

---

[111] Dalton et al., 1980; Daft, 2008, p. 15; Pugh et al., 1968; Henry Mintzberg, "Structure in 5's: A Synthesis of the Research on Organization Design," *Management Science* 26, no. 3 (1980).

[112] Blau, 1968; Mintzberg, 1980.

[113] Dalton et al., 1980.

Appendix D. InfoSec Suborganizations

How to organize InfoSec teams is currently an active research topic, with academic literature, business literature, and industry surveys tackling the subject. This likely reflects the fact that InfoSec is a rapidly evolving field with considerably more commercial sector and public sector emphasis than it had in the past.

Forrester Research, a global research and advisory firm, released a report in 2010 intended "to help security and risk professionals build an effective [information] security organization."[114] The report was based on analyst experience and a survey of more than 2,000 companies across many industry sectors. The report describes a model InfoSec organization that we summarize here using NICE terminology. In the report's construct, a CISO led the InfoSec organization, which was primarily composed of four groups—security oversight, IT risk, security engineering, and security operations, depicted in Figure D.1. Our interviews revealed differences in how these subgroups are combined, but the companies we interviewed reported performing similar functions and missions.

The security oversight group contains program management, business liaisons, metrics and reporting, and marketing. Specific tasks include advocating for the unique needs of the business units within the security team, relaying information about the value-added of the security team to the business as a whole, and coordinating between the security team and business units. This group encompasses functions similar to those in the NICE Framework's *oversight and development* category.

The IT risk group is concerned with IT policy and compliance standards, monitoring third-party security, managing threat and vulnerabilities, and conducting risk assessments. Specific tasks include ensuring that data shared with outside vendors are protected, communicating threats to the wider organization to raise awareness, and prioritizing security activities. This group performs functions most closely aligned with the NICE Framework *oversight and development* functions related to information systems security operations, although it also contains elements of *analyze* with regards to threat, as well as *protect and defend* in vulnerability management.

The security engineering group establishes system policies and architecture, enforces application security, implements security measures, and integrates security tools. Specific tasks include translating policy into lower-level guidance and testing third-party applications. This group most closely parallels the *securely provision* functions established in the NICE

[114] Kark and Dines, 2010.

Framework, although it also comprises the computer network defense infrastructure support task for the *protect and defend* function.

Figure D.1. InfoSec Organization in NICE Terminology

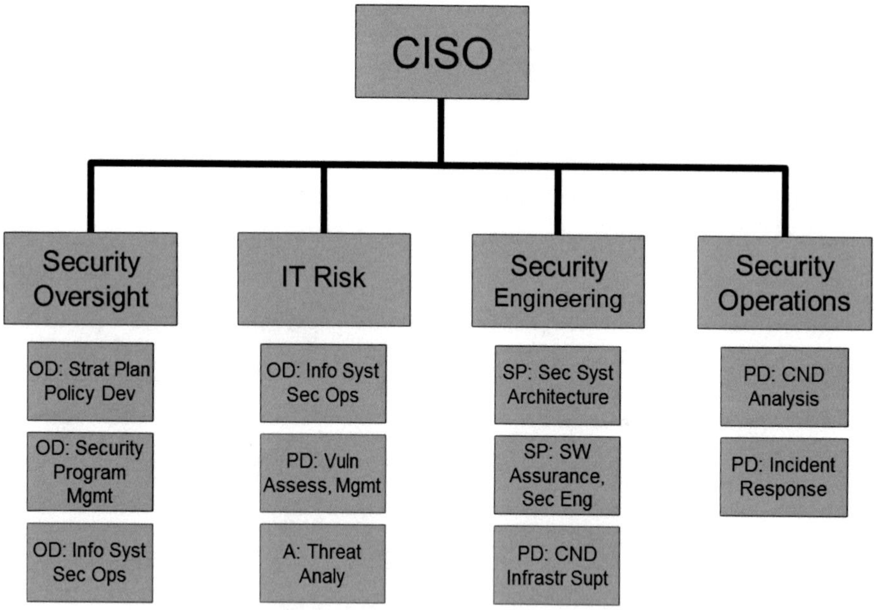

SOURCE: Analysis of Kark and Dines, 2010, p. 8.

NOTE: OD = oversight and development; Strat Plan = strategic planning; Dev = development; Mgmt = management; Info Syst = information system; Sec = security; Ops = operations; PD = protect and defend; Vuln Assess, Mgmt = vulnerability assessment and management; Analy = analysis; SP = securely provision; SW = software; Eng = engineering; CND = computer network defense; Infrastr Supt = infrastructure support.

Finally, the security operations group is responsible for the infrastructure security, monitoring of devices, management of security information, and response to security incidents. Specific tasks include ensuring that servers, routers, and workstations are securely configured and correlating disparate information gathered throughout the system to detect intrusions. This group most closely parallels *protect and defend* functions established in the NICE Framework.

Next we examine one particular model for a computer incident response team. As a specialized subset of InfoSec, some literature focused on the organization of CERTs. The Internet Engineering Task Force sets out in detail a set of activities and tasks that a CERT should accomplish, along with suggested guidance on what timescales are appropriate to react to certain events. [115] The services specified include incident triage, incident coordination, and incident resolution. Industry publications further make recommendations for the best way to organize a

---

[115] N. Brownlee and E. Guttman, *Request for Comments 2350 Expectations of Computer Security Incident Response*, Network Working Group, The Internet Society, 1998.

CERT. Experts from a well-known cybersecurity firm advocate a CERT composed of three divisions and a constituent relations team, shown in Figure D.2.[116]

Figure D.2. CERT Organization

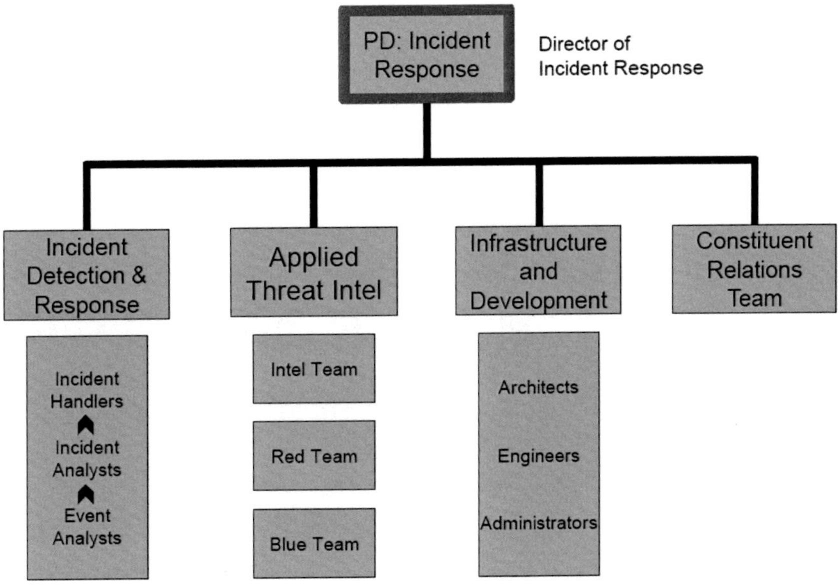

SOURCE: Analysis of Bejtlich, 2013.

In this CERT design, the first division is for incident detection and response, which most closely parallels the functions of *protect and defend* from the NICE Framework, is required to provide response capabilities 24 hours a day, seven days a week. This division is composed of incident handlers, who are experienced analysts tasked with finding intruders on the network (also known as hunting); incident analysts, who are midlevel analysts tasked with finding intruders either by matching known indicators (e.g., malicious IP addresses or domain names) or using hunting techniques; and event analysts, who are junior analysts primarily focused on finding intruders by matching known indicators. Typically, event analysts receive on-the-job training from the incident analysts and incident handlers, and over the course of years, event analysts rise through the three levels as they develop their creativity and critical thinking skills.

Second, an applied threat intelligence division parallels the analyze function from the NICE Framework. The applied threat intelligence division is composed of principal, senior, and associate analysts tasked with intelligence activities, penetration testing of the company's network, adversary simulation (a.k.a., red teaming), and internal security consulting (a.k.a., blue teaming). The subgroups of this division are clearly mission-oriented.

---

[116] Richard Bejtlich, *The Practice of Network Security Monitoring: Understanding Incident Detection and Response*, San Francisco, Calif.: No Starch Press, 2013.

Third, an infrastructure and development division is composed of network architects, software engineers, and administrators performing some of the functions of *securely provision* from the NICE Framework. They help design the system's security architecture, develop production-grade tools, and lead the development of new detection and response techniques. This is a cross-functional division, bringing in a wide variety of skills and abilities.

Fourth, a constituent relations team is responsible for liaising with other parts of the organization to make sure the CERT is able to work effectively across the entire organization and understands the nuances of each business unit and division. The European Union's cyber security agency, the European Network and Information Security Agency, suggests that the cooperation and liaison functions of CERTs should also include the ability to liaise with other stakeholders *outside* the organization, including peers and other types of CERTs in different countries. In a similar vein, there is increasing understanding that legal knowledge (often held by a manager or team leader) is an important part of incident response, especially regarding the legal permissibility of actions.[117] Tied to this, larger incident response teams are also increasingly recognizing the importance of having access to law enforcement knowledge, either as dual-hatted personnel or via dual-trained staff.[118]

---

[117] European Network and Information Security Agency, *A Flair for Sharing—Encouraging Information Exchange Between CERTs*, Heraklion, Greece, 2011.

[118] European Network and Information Security Agency, *Cooperation Between CERTs and Law Enforcement Agencies in the Fight Against Cybercrime—A First Collection of Practices*, Heraklion, Greece, 2012.

References

ADP, *Effective Talent Management Has Become an Essential Strategy for Organizational Success*, Alpharetta, Ga.: ADP, Inc., 2010.

Agile Methodology, website, undated. As of November 11, 2014: http://agilemethodology.org/

Air Force Personnel Center, "Air Force Personnel Demographics," March 31, 2014. As of July 7, 2014: http://www.afpc.af.mil/library/airforcepersonneldemographics.asp

Bejtlich, Richard, *The Practice of Network Security Monitoring: Understanding Incident Detection and Response*, San Francisco, Calif.: No Starch Press, 2013.

Bell, Gerald D., "Determinants of Span of Control," *American Journal of Sociology* 73, no. 1 (1967): 100–109.

Bishop, Matt, "Teaching Context in Information Security," *ACM Journal of Educational Resources in Computing* 6, no. 3 (September 2006).

Blau, Peter M., "The Hierarchy of Authority in Organizations," *American Journal of Sociology* 73, no. 4 (1968): 453–467.

Bohte, John, and Kenneth J. Meier, "Structure and Performance of Public Organizations: Task Difficulty and Span of Control," *Public Organization Review* 1, no. 3 (2001): 341–354.

Booz Allen Hamilton, *Cyber Training: Developing the Next Generation of Cyber Analysts*, 2011. As of September 23, 2014: http://www.boozallen.com/media/file/Cyber_Human_Capital.pdf

Brownlee, N., and E. Guttman, *Request for Comments 2350 Expectations of Computer Security Incident Response*, Network Working Group, The Internet Society, 1998. As of July 10, 2014: http://www.ietf.org/rfc/rfc2350.txt

Bureau of Labor Statistics, "Computer and Information Technology Occupations," in *Occupational Outlook Handbook*, Washington, D.C., January 8, 2014a. As of July 5, 2014: http://www.bls.gov/ooh/computer-and-information-technology/home.htm

———, "Network and Computer Systems Administrators: How to Become a Network and Computer Systems Administrator," in *Occupational Outlook Handbook*, Washington, D.C., January 8, 2014b. As of July 5, 2014: http://www.bls.gov/ooh/computer-and-information-technology/network-and-computer-systems-administrators.htm#tab-4

———, *Occupational Outlook Handbook*, Washington, D.C., January 8, 2014c. As of July 8, 2014: http://www.bls.gov/ooh/

Business Editors, *Survey Indicates No "One Size Fits All" Solution to IT Structures and Staffing; Joint Study Released by people3, Mercer Human Resource Consulting and ITAA*, Business Wire, February 3, 2003. As of May 1, 2014: http://www.thefreelibrary.com/Survey+Indicates+No+%27One+Size+Fits+All%27+Solution+to+IT+Structures+and...-a097180235

Cezar, Asunur, Huseyin Cavusoglu, and Srinivasan Raghunathan, "Outsourcing Information Security: Contracting Issues and Security Implications," *Management Science* 60, no. 3 (2014): 638–657.

Chang, Young Bong, and Vijay Gurbaxani, "Information Technology Outsourcing, Knowledge Transfer, and Firm Productivity: An Empirical Analysis," *MIS Quarterly* 36, no. 4 (2012): 1043–1063.

Chen, Xianglei, and Matthew Soldner, *STEM Attrition: College Students' Paths into and out of STEM Fields: Statistical Analysis Report*, Washington, D.C.: U.S. Department of Education, November 2013. As of July 16, 2014: http://nces.ed.gov/pubs2014/2014001rev.pdf

Conti, Gregory, and David Raymond, "Leadership of Cyber Warriors: Enduring Principles and New Directions," *Small Wars Journal*, July 11, 2011. As of July 8, 2014: http://smallwarsjournal.com/jrnl/art/leadership-of-cyber-warriors-enduring-principles-and-new-directions

Daft, Richard L., *Organization Theory and Design*, 10th ed., Mason, Ohio: South-Western Cengage Learning, 2008.

Dalton, Dan R., William D. Todor, Michael J. Spendolini, Gordon J. Fielding, and Lyman W. Porter, "Organization Structure and Performance: A Critical Review," *Academy of Management Review* 5, no. 1 (1980): 49–64.

Dodge, Ronald C., Costis Toregas, and Lance Hoffman, "Cybersecurity Workforce Development Directions," in *Proceedings of the Sixth International Symposium on Human Aspects of Information Security & Assurance: HAISA 2012*, ed. Nathan Clarke and Steven Furnell, Plymouth, UK: University of Plymouth, 2012.

Duncan, Robert, "What Is the Right Organization Structure? Decision Tree Analysis Provides the Answer," *Organizational Dynamics* 79, no. 3 (1979): 59–80.

Economist Intelligence Unit, *Business Resilience: Ensuring Continuity in a Volatile Environment, Economist Insights*, London: The Economist, 2007. As of May 7, 2014: http://www.economistinsights.com/sites/default/files/eiu_Bus_Resilience_wp.pdf

Ernst & Young, *Fighting to Close the Gap: Global Information Security Survey 2012*, November 2012.

European Network and Information Security Agency, *A Flair for Sharing—Encouraging Information Exchange Between CERTs*, Heraklion, Greece, 2011. As of July 10, 2014: https://www.enisa.europa.eu/activities/cert/support/fight-against-cybercrime/legal-information-sharing

———, *Cooperation Between CERTs and Law Enforcement Agencies in the Fight Against Cybercrime—A First Collection of Practices*, Heraklion, Greece, 2012. As of July 10, 2014: https://www.enisa.europa.eu/activities/cert/support/fight-against-cybercrime/supporting-fight-against-cybercrime

Fortune, "Global 500 2014," undated. As of September 23, 2014: http://fortune.com/global500/

Fukuyama, Francis, and Abram N. Shulsky, *The "Virtual Corporation" and Army Organization*, Santa Monica, Calif.: RAND Corporation, MR-863-A, 1997. As of September 23, 2014: http://www.rand.org/pubs/monograph_reports/MR863.html

Galbraith, Jay, "Matrix Organization Designs: How to Combine Functional and Project Forms," *Business Horizons* 14, no. 1 (1971): 29–40.

———, *Designing Complex Organizations*, Reading, Mass.: Addison-Wesley, 1973.

GoCertify, "Most Popular IT Certifications," web page, undated. As of July 5, 2014: http://www.gocertify.com/certifications/index.html

Goel, Sanjay, Damira Pon, Peter Bloniarz, Robert Bangert-Drowns, George Berg, Vince Delio, Laura Iwan, Thomas Hurbanek, Sandoor P. Schuman, Jagdish Gangolly, Adnan Baykal, and Jon Hobbs, "Innovative Model for Information Assurance Curriculum: A Teaching Hospital," *ACM Journal on Educational Resources in Computing* 6, no. 3 (September 2006).

Gonsalves, Antone, "Target Top Security Officer Reporting to CIO Seen as Mistake," *CSO Online*, June 13, 2014. As of July 14, 2014: http://www.csoonline.com/article/2363210/data-protection/target-top-security-officer-reporting-to-cio-seen-as-a-mistake.html

Grasso, Valerie Bailey, *Defense Outsourcing: The OMB Circular A-76 Policy*, Washington, D.C.: Congressional Research Service, 2005.

Hackett Group, "Hackett: Companies Can Improve Earnings Nearly 15% by Improving Talent Management Function," July 24, 2007. As of July 5, 2014: http://www.thehackettgroup.com/about/alerts/alerts_2007/alert_07242007.jsp

Han, Kunsoo, and Sunil Mithas, "IT Outsourcing and Non-IT Operating Costs: An Empirical Investigation," *MIS Quarterly* 37, no. 1 (2013): 315–331.

Harrell, Margaret C., Harry J. Thie, Roland J. Yardley, and Maria C. Lytell, *Information Systems Technician Rating Stakeholders: Implications for Effective Performance*, Santa Monica, Calif.: RAND Corporation, TR-1122-NAVY, 2011. As of September 23, 2014: http://www.rand.org/pubs/technical_reports/TR1122.html

Hausknecht, John P., Julianne Rodda, and Michael J. Howard, "Targeted Employee Retention: Performance-Based and Job-Related Differences in Reported Reasons for Staying," *Human Resource Management* 48, no. 2 (2009): 269–288.

HDI, *2013 Support Center Practices and Salary Report*, Colorado Springs, Colo.: HDI, 2013.

Hoffman, Lance J., Tim Rosenberg, Ronald Dodge, and Daniel Ragsdale, "Exploring a National Cybersecurity Exercise for Universities," *IEEE Security & Privacy* 3, no. 5 (2005): 27–33.

Homan, Timothy, and Zachary Tracer, "ADP Estimates Companies in U.S. Added 42,000 Jobs," *Bloomberg*, August 4, 2010. As of July 7, 2014: http://www.bloomberg.com/news/2010-08-04/u-s-companies-added-more-than-forecast-42-000-jobs-last-month-adp-says.html

Indeed, homepage, undated. As of July 5, 2014: http://www.indeed.com

$(ISC)^2$, "CISSP—Certified Information Systems Security Professional," undated. As of May 6, 2014: https://www.isc2.org/CISSP/Default.aspx

Jaques, Elliott, "In Praise of Hierarchy," *Harvard Business Review* (January–February 1990): 127–133.

Joint Publication 1-02, *Department of Defense Dictionary of Military and Associated Terms*, Washington, D.C.: Department of Defense, November 8, 2010, as amended through August 15, 2014. As of July 8, 2014: http://www.dtic.mil/doctrine/new_pubs/jp1_02.pdf

Joint Task Force for Computing Curricula, *Computing Curricula 2005: The Overview Report*, Cambridge, Mass.: Association for Computing Machinery and Institute of Electrical and Electronics Engineers, 2006.

———, *Computer Science Curricula 2013*, Association for Computing Machinery and Institute of Electrical and Electronics Engineers, 2013.

Kaplan, James, Naufal Khan, and Roger Roberts, "Winning the Battle for Technology Talent," McKinsey & Company, May 2012. As of July 8, 2014: http://www.mckinsey.com/insights/business_technology/winning_the_battle_for_technology_talent

Kark, Khalid, and Rachel A. Dines, *Security Organization 2.0: Building a Robust Security Organization*, Cambridge, Mass.: Forrester Research, Inc., 2010.

Killcrece, Georgia, Klaus-Peter Kossakowski, Robin Ruefle, and Mark Zajicek, *Organizational Models for Computer Security Incident Response Teams (CSIRTs)*, Pittsburgh, Pa.: Software Engineering Institute, Carnegie Mellon University, 2003.

Libicki, Martin C., Dave Senty, and Julia Pollak, *Hackers Wanted: An Examination of the Cybersecurity Labor Market*, Santa Monica, Calif.: RAND Corporation, RR-430, 2014. As of September 23, 2014: http://www.rand.org/pubs/research_reports/RR430.html

Lockheed Martin, "Lockheed Martin Center for Security Analysis (LMCSA)," undated. As of July 5, 2014: http://www.lockheedmartin.com/content/dam/lockheed/data/isgs/documents/LMCSA%20Overview%20Fact%20Sheet_Final%20%282%29.pdf

Lunt, Barry, Joseph J. Ekstrom, Sandra Gorka, Gregory Hislop, Reza Kamali, Eydie Lawson, Richard LeBlanc, Jacob Miller, and Han Reichgelt, *Information Technology 2008: Curriculum Guidelines for Undergraduate Degree Programs in Information Technology*, Cambridge, Mass.: Association for Computing Machinery and Institute of Electrical and Electronics Engineers, 2008.

Mandiant, *APT1: Exposing One of China's Cyber Espionage Units*, February 2013.

McCullough, Amy, "Cyber Futures," *Air Force Magazine*, June 2011, pp. 34–39.

Microsoft, "MCSE: Server Infrastructure," undated-a. As of May 6, 2014: http://www.microsoft.com/learning/en-us/mcse-server-infrastructure-certification.aspx

———, "MCSA: Windows Server, 2012," undated-b. As of May 6, 2014: http://www.microsoft.com/learning/en-us/mcsa-windows-server-certification.aspx

———, "Microsoft Technology Associate (MTA)," undated-c. As of May 6, 2014: http://www.microsoft.com/learning/en-us/mta-certification.aspx

Mintzberg, Henry, "Structure in 5's: A Synthesis of the Research on Organization Design," *Management Science* 26, no. 3 (1980): 322–341.

———, "Organization Design: Fashion or Fit?" *Harvard Business Review* 59, no. 1 (1981): 103–116.

MITRE, "Cybersecurity Awareness and Training," 2014. As of July 5, 2014: http://www.mitre.org/capabilities/cybersecurity/overview/cybersecurity-resources/awareness-training

National Cyber Watch Center, homepage, undated. As of July 5, 2014: http://www.cyberwatchcenter.org/

National Initiative for Cybersecurity Careers and Studies, *Interactive National Cybersecurity Workforce Framework*, Washington, D.C.: Department of Homeland Security, undated. As of July 2, 2014: http://niccs.us-cert.gov/training/tc/framework

National Initiative for Cybersecurity Education. *National Cybersecurity Workforce Framework*, Washington, D.C.: Department of Commerce, 2013. As of July 2, 2014: http://csrc.nist.gov/nice/framework/

National Institute of Standards and Technology, *Information Security Handbook: A Guide for Managers*, Special Publication 800-100, Gaithersburg, Md., October 2006.

National Research Council, *Building a Workforce for the Information Economy*, Washington, D.C.: The National Academies Press, 2001.

———, *Professionalizing the Nation's Cybersecurity Workforce? Criteria for Decision-Making*, Washington, D.C.: The National Academies Press, 2013.

National Security Agency and Central Security Service, "National Centers of Academic Excellence in Information Assurance (IA)/Cyber Defense (CD)," posted on January 15, 2009, last modified August 20, 2014. As of September 23, 2014: http://www.nsa.gov/ia/academic_outreach/nat_cae/index.shtml

Northrop Grumman, "Cyber Academy: Developing the Cyber Workforce," 2013. As of July 5, 2014: http://www.northropgrumman.com/Capabilities/Cybersecurity/Documents/CyberAcademy_overview.pdf

Office of the Under Secretary of Defense for Personnel and Readiness, "Regular Military Compensation Calculator," undated. As of January 19, 2015: http://militarypay.defense.gov/pay/calc/index.html

OpenSecurityTraining.info, homepage, undated. As of July 5, 2014: http://opensecuritytraining.info/

Oracle, *Talent Retention: Six Technology-Enabled Best Practices*, Oracle Corporation, 2012.

Paul, Christopher, Harry J. Thie, Katharine Watkins Webb, Stephanie Young, Colin P. Clarke, Susan G. Straus, Joya Laha, Christine Osowski, and Chad C. Serena, *Alert and Ready: An Organizational Design Assessment of Marine Corps Intelligence*, Santa Monica, Calif.: RAND Corporation, MG-1108-USMC, 2011. As of September 23, 2014: http://www.rand.org/pubs/monographs/MG1108.html

Pike, Ronald, "The Case for Depth in Cybersecurity Education," *ACM Inroads* 5, no. 1 (2014): 47–52.

Prahalad, C. K., and Gary Hamel, "The Core Competence of the Corporation," *Harvard Business Review* (May–June 1990): 79–90.

Pugh, D. S., D. J. Hickson, C. R. Hinings, and C. Turner, "Dimensions of Organization Structure," *Administrative Science Quarterly* 13, no. 1 (1968): 65–105.

Raytheon, *Cyber Learning Solutions: 2012–2013 Extended Course Catalog*, 2012. As of July 7, 2014: http://www.raytheon.com/capabilities/rtnwcm/groups/public/documents/content/rtn_131003.pdf

Schooley, Claire, Connie Moore, and Ralph Vitti, *Forrester Research: Drive Employee Talent Development Through Business Mentoring Programs*, Cambridge, Mass.: Forrester Research, Inc., 2010.

Snyder, Don, Bernard Fox, Kristin F. Lynch, Raymond E. Conley, John A. Ausink, Laura Werber, William Shelton, Sarah A. Nowak, Michael R. Thirtle, and Albert A. Robbert, *Assessment of the Air Force Materiel Command Reorganization: Report for Congress*, Santa Monica, Calif.: RAND Corporation, RR-389-AF, 2013. As of September 23, 2014: http://www.rand.org/pubs/research_reports/RR389.html

Stark, J., M. Arlt, and D. H. T. Walker, "Outsourcing Decisions and Models—Some Practical Considerations for Large Organizations," in *International Conference on Global Software Engineering, 2006: ICGSE '06*, 12–17, Los Alamitos, Calif.: IEEE Computer Society Press, 2006.

Suby, Michael, *The 2013 (ISC)^2 Global Information Security Workforce Study*, Mountain View, Calif.: Frost & Sullivan, 2013.

Thales Group, "Thales Unveils Cyber Integration & Innovation Centre," October 24, 2013. As of July 2, 2014: https://www.thalesgroup.com/en/cybersecurity/press-release/thales-unveils-cyber-integration-innovation-centre

Theoharidou, Marianthi, and Dimitris Gritzalis, "Common Body of Knowledge for Information Security," *IEEE Security & Privacy* 5, no. 2 (2007): 64–67.

Tobey, David H., Portia Pusey, and Diana L. Burley, "Engaging Learners in Cybersecurity Careers: Lessons from the Launch of the National Cyber League," *ACM Inroads* 5, no. 1 (March 2014): pp. 53–56.

U.S. General Accounting Office, *Information Technology: DOD Needs to Leverage Lessons Learned from Its Outsourcing Projects*, Washington, D.C., 2003.

———, *Information Technology: DOD Needs to Ensure That Navy Marine Corps Intranet Program Is Meeting Goals and Satisfying Customers*, Washington, D.C., 2006.

Verizon RISK Team, *2013 Data Breach Investigations Report*, 2013. As of July 7, 2014: http://www.verizonenterprise.com/resources/reports/rp_data-breach-investigations-report-2013_en_xg.pdf

von Simon, Ernest M., "The 'Centrally Decentralized' IS Organization," *Harvard Business Review* (July–August 1990): 158–162.

White House, *The Clinton Administration's Policy on Critical Infrastructure Protection: Presidential Decision Directive 63*, white paper, Washington, D.C., 1998.

Willem, Annick, and Marc Buelens, "Knowledge Sharing in Public Sector Organizations: The Effect of Organizational Characteristics on Interdepartmental Knowledge Sharing," *Journal of Public Administration Research and Theory* 17, no. 4 (2007): 581–606.

Workforce.com, "Ratio of IT Staff to Employees," February 6, 2003. As of July 8, 2014: http://www.workforce.com/articles/ratio-of-it-staff-to-employees

MUSIC ALIVE!™
Music Education for Today's Generation

Music from Around the World

By Brad Shank

Editorial Director: Caroline Horn
Compilation Editor: Carolyn Keating
Copy Editor and Production Manager: Susanah Doucet
Art Direction: Doreen Walsh
Cover Design: José Ortega
Cartography: Jim McMahon
Music Transcription: Adam Perlmutter
Editorial Interns: Marquita Benton, Antonia Cove

Published by
Cherry Lane Music Foundation, Inc.
6 East 32nd Street, 11th Floor
New York, NY 10016
www.cherrylane.com
www.musicalive.com

cherry lane
music foundation, inc.

Exclusively Distributed By
HAL•LEONARD®
CORPORATION
7777 W. Bluemound Rd. P.O. Box 13819 Milwaukee, WI 53213

ISBN: 978-1-60378-008-7

Contents

About the Author

Brad Shank received his Master of Arts in Ethnomusicology from the University of California, Los Angeles, where he studied world music styles and cultures with a focus on African popular music and American folk music. As a music educator, he has taught world music curriculum and performance at all levels, from elementary to collegiate, and has presented at conferences hosted by MENC, the national association of music educators.

With the goal of making world music more approachable, interesting, and exciting to students and teachers, Brad has written extensively on music styles and performers from around the globe. As supervising editor in the Music Division at Macmillan/McGraw-Hill, he helped develop the K–8 music program *Spotlight on Music*, and as a contributor to the elementary level music education magazine *Music Express!*, he researched folk songs and created lesson plans. He also collaborated with ethnomusicologist John Higgins on *More Music of Our World*, a collection of songs, articles, and activities with global origins.

Brad has been a writer and editor for *Music Alive!* magazine for many years. He is currently an executive editor with McGraw-Hill's School Solutions Group.

Music Alive! wishes to thank the following record labels for their generous contributions of the music recordings for this classroom volume:

ARC Music International
Gallo Records
Ice Records, Ltd.
Music Today
Rounder Records
Shanachie Entertainment
Smithsonian Folkways

A Note on Terms

We have used a variety of styles to signify the different types of musical terms and special words found in our feature articles:

WORDS IN SMALL CAPITAL LETTERS are select vocabulary words that are defined in the corresponding lesson plan.

Words in italics are foreign words.

Words in boldface are terms that are defined in the glossary at the back of the book.

Words in both boldface and italics are foreign terms that are also defined in the glossary.

Introduction

Bent over, straining to read in the dark light between the rows of bookshelves, I got my introduction to world music via a serendipitous discovery late one night at my college music library. In the middle of some research, I quite suddenly came across a CD of *mbira* (em-BEE-rah) music from Zimbabwe. The picture of the *mbira* on the cover, from this small country in southern Africa, intrigued me. Its rows of spoon-like keys positioned on a rectangular wooden base were like nothing I had seen before.

So I paused my research to pop the disc into a player at the library. Bell-like splashes of sound bubbled in a playful, rhythmic mosaic through the earphones. In an instant, I understood there might be a huge world of music beyond the study of Western classical music, which had been my focus. Thus began a life-long interest in world music that included a semester studying in Zimbabwe—and led me to a deeper understanding not only of music, but of people, cultures, and our place in the world.

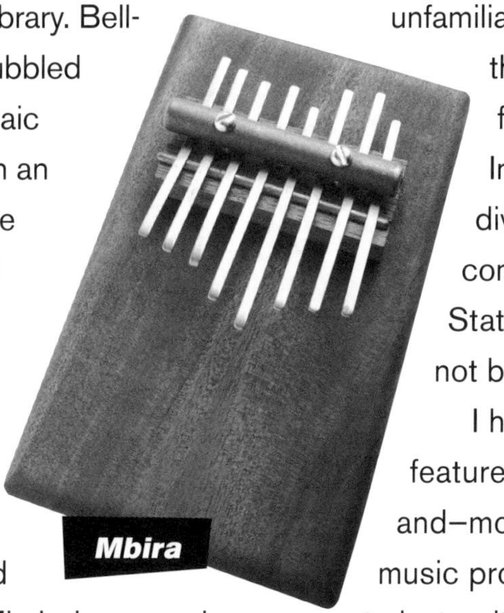

Mbira

Like me, you and your students can discover new sounds beyond the limits of your imagination. Meanwhile, you can harness the power of music to motivate, captivate, and stimulate students to learn. As a music teacher, you already know that music fosters retention, encourages focus, and, of course, offers loads of fun.

But it can do so much more. I have always found music to be a wonderful entry into new cultures. Catch their attention with great music, and soon students will be eager to learn about a country, its history, and its people. And teaching students to respect unfamiliar music styles helps them to learn respect for unfamiliar cultures. In the increasingly diverse classrooms and communities of the United States, this revelation could not be more important!

I hope this collection of feature articles, lesson plans, and—most importantly—terrific music provides you and your students with an opportunity to begin (or continue) down a fulfilling path full of discovery and appreciation for music from around the world. **—Brad Shank**

Sounds of South Africa

South Africa spreads across the entire southern tip of Africa, taking up an area larger than California and Texas combined. The earliest inhabitants of South Africa were probably ancestors of the San and Khoi peoples. (One example of their influence is the "click" consonants found in several South African languages, including Xhosa, Sotho, and Zulu.) Later, several waves of immigrants came to South Africa from central Africa.

In 1652, Dutch explorers landed on the shores of South Africa. French and German settlers arrived a little while later and mixed with the Dutch, and this group of people became known as "Afrikaners." By the early 1800s, people from England were arriving, followed by immigrants from India, China, and other parts of Asia.

This blend of cultures produced an unfortunate and TURBULENT history in South Africa. When the European settlers arrived, they pushed the native people out of their original homelands. Then, in the mid-1900s, white rulers created an OPPRESSIVE system of government called **apartheid.** Black South Africans were forced to move into COMMUNAL areas in the country and overcrowded areas called "townships" in the cities; they couldn't leave these areas without special passes. In the 1960s and 1970s, a struggle led by people like Nelson Mandela and Stephen Biko gathered strength to end apartheid. After so many years of hardship, apartheid ended in 1994. An African National Congress was in place and black South Africans were finally free.

Deep Roots

Despite the troubles the country has known, South Africa has produced very joyous music styles. There have always been wonderful vocal harmonies, beginning with the people of the San tribe. Later, immigrants from other African countries brought instruments such as drums, ***mbiras*** (thumb pianos), and **marimbas** to the area. European settlers introduced the guitar, the accordion, and the penny whistle. Many black South Africans were drawn to European church music,

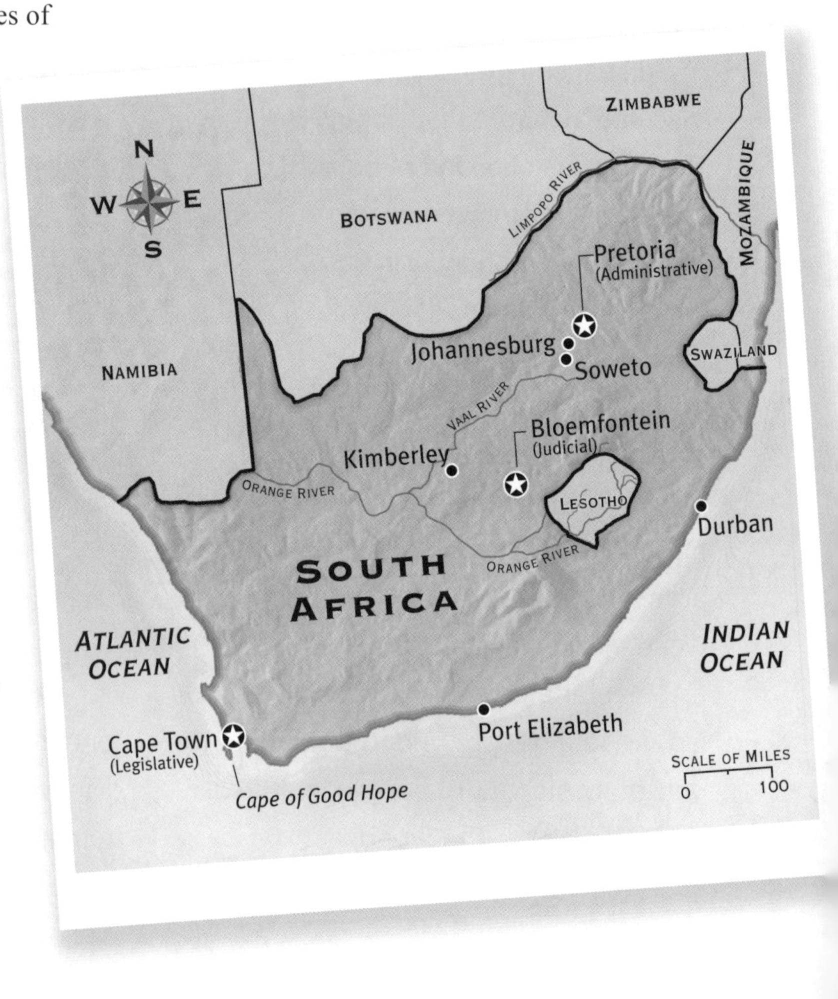

which they adopted and quickly made their own.

The musical connection between South Africa and the United States dates back over 100 years. In the late 1800s, several groups of African-American minstrel singers from the United States traveled to South Africa, where they performed spirituals, folk songs, and other popular music of the time. This choral sound is the backbone of many South African styles. For example, ***isicathimiya*** (iss-ee-kah-tah-MEE-yah) is a popular *a cappella* style of usually all-male choirs in which a solo singer calls to other choir members who respond with a low, rich harmony. Ladysmith Black Mambazo, led by Joseph Shabalala, is the most famous *isicathimiya* group in South Africa.

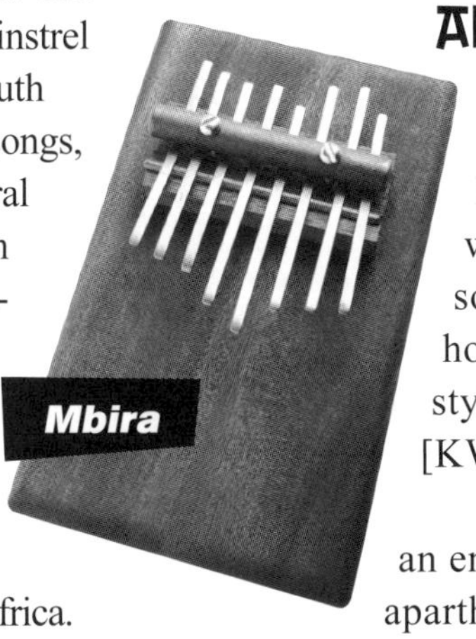

Mbira

In the 1960s, the traditional Zulu song "Mbube (The Lion)" also brought the sounds of South Africa to American ears when it was rearranged by Pete Seeger and the Weavers as "Wimoweh (The Lion Sleeps Tonight)."

Retro in Soweto

In the United States, we recognize old or "retro" styles like jump blues and rockabilly as important pieces of our musical history. The same is true in South Africa. The pianos, banjos, and guitars of ***marabi*** bands had people swinging throughout the country in the 1930s. Using penny whistles to improvise jazzy tunes, ***kwela*** musicians like Spokes Mashiyane hypnotized people all across Africa in the 1950s. Later, the ***mbaqanga*** style replaced the penny whistle with a sax and conquered the South African scene. Simon "Mahlathini" Nkabinde and the Mahotella Queens ENTHRALLED audiences with a sound called "Zulu township **jive**."

The grandparents of modern South African music are vocalist Miriam Makeba and jazz trumpeter Hugh Masekela. Makeba became famous with her version of the "Click Song"—originally used to teach children the traditional Xhosa language—as well as "Pata Pata"; both became worldwide hits. Masekela studied in London and also lived in New York City before returning to South Africa. In 1968, his instrumental composition "Grazing in the Grass" topped the Rolling Stones' "Jumpin' Jack Flash" to reach #1 on the U.S. pop charts!

All Their Own

If you visit South Africa today, you will find one of the hottest music scenes in the world. Kids in Johannesburg, Durban, or Cape Town will introduce you to the cool beats and slick sounds they hear daily. South African hip-hop blasts from speakers everywhere, but one style captures young South Africa best: ***kwaito*** [KWEYE-toe].

Born from the ashes of apartheid, *kwaito* is an entirely original South African creation. Once apartheid ended, young people and musicians were finally able to say and play whatever they wanted. Inspired by this new feeling of freedom, many musicians created a whole new style, mixing American hip-hop and house beats with local song styles like *marabi* and *kwela,* and even Jamaican reggae and dancehall. Like hip-hop, *kwaito* lyrics tell stories that young South Africans can relate to.

Many *kwaito* groups have begun to add live instruments to the vibe. The Brothers of Peace and Bongo Maffin are two groups demonstrating this trend.

Free and Flowing

After a little over a decade of complete musical freedom, South African musicians are just getting warmed up. "We are taking the best from overseas, but then doing our own thing. There is a real RENAISSANCE going on," said Hugh Masekela. Who knows what this creative country full of eager and talented musicians will serve up next? ◎

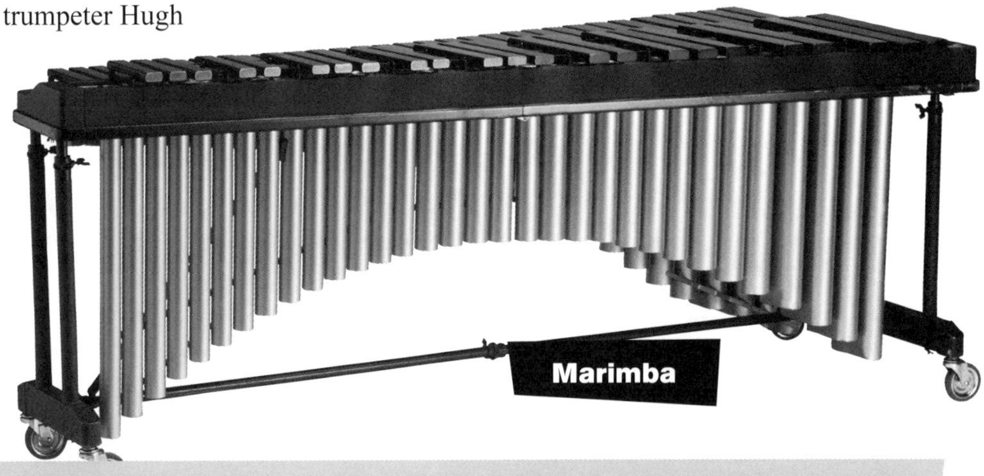

Marimba

◎ listen to: POP-MBAQANGA FROM THE MAHOTELLA QUEENS AND ISICATHIMIYA FROM THE SOWETO GOSPEL CHOIR

LESSON PLAN

Sounds of South Africa

Objectives

- Discuss the term "world music" (NS 9)
- Listen to and analyze two songs from South Africa (NS 6)
- Perform rhythm patterns with a South African song (NS 2, 5)
- Create a call-and-response verse (NS 4)

Materials

- *Music from Around the World* CD
- Congas or other classroom drums
- Student Worksheet: Call and Response (page 10)

START

1 People who live in the United States often have a limited perspective of music outside of their country. Lead a short discussion about the term "world music" to encourage consideration of these issues.

ASK **Have you ever seen the genre "world music" in a music store? What do you think world music is?** (Answers will vary.)
Do you think all world music sounds alike? Why or why not? (Answers will vary.)

Explain that the term "world music" is hard to define because it encompasses such a large range of styles from many different countries. Share that sometimes people use the terms "folk" and "popular" to describe music from other countries. Ask students why they think that is. (Because other countries have different genres of music, just as we do.)

2 Divide the class into pairs. Have each pair take five minutes to create a list of as many different music genres as they can think of. (Examples can include rock, pop, classical, blues, rap, hip-hop, jazz, metal, etc.) Then ask volunteers to call out different genres while you write these on the board.

ASK **Do you think that categorizing all music in the United States as "rock" or "classical" gives the rest of the world an accurate picture of American music? Why or why not?**

DEVELOP

1 **LITLINK** Have students split up into pairs and take turns reading "Sounds of South Africa" (pp. 6–7) aloud. Review the highlighted vocabulary words as a class and call on different students to determine the meaning of each word from its context. **[This reinforces vocabulary.]**

TURBULENT—characterized by unrest, violence, agitation
OPPRESSIVE—unreasonably or cruelly overpowering
COMMUNAL—shared, used by a group
ENTHRALLED—charmed or spellbound
RENAISSANCE—rebirth

2 **LITLINK** Check students' comprehension of the article by having each pair answer the following questions on a separate sheet of paper. **[This reinforces comprehension.]**

ASK **What is a new style of music in South Africa that started after the end of apartheid?** *(Kwaito)* **Is it similar to any music in the United States?** (Yes, hip-hop)
In what ways did many people in South Africa suffer under apartheid? (They lived in poor areas; they couldn't go where they wanted or express themselves fully.) **Do you think this affected the music they created? Why or why not?**

2 Tell students that they will now hear two different styles of music from South Africa: a pop style and a folk style. Play "Women of the World" (CD track 1). Ask students to call out what instruments they heard in this song. Then ask them if they think any of these instruments came from outside South Africa. (All did—guitar, bass, drum kit, synthesizer, and penny whistle are not South African.) Point out that this foreign instrumentation shows how outside styles have influenced South African popular music.

3 **LITLINK** Now play "Thina Simnqobile" (CD track 2). Tell students that the style of this song is mentioned in the article; challenge them to scan the article and guess which style it is (it is *isicathimiya*). **[This reinforces comprehension.]**

Ask students to listen for the low male bass harmonies common to *isicathimiya* that can be heard in the beginning *a cappella* section (0:00–0:50).

4 Next, have students listen to the group sing the title of the song at 0:50.

ASK **Do you notice anything different in the way they pronounce the title?** (They click their tongues.)

Explain that several languages in South Africa, such as Zulu, feature clicking syllables. The "q" in the word *simnqobile* represents a special click sound that is made by popping the tongue on the roof of the mouth. The click can be heard beginning at 0:50 when the vocalists begin singing the title.

5 Distribute copies of the Call and Response worksheet. Review any songs in call and response (or solo/chorus) form that students may have sung. Have students complete the worksheet. If they need additional guidance, provide the following example:

Setting: A sporting event
Call: Who is going to win this game?
Response: We are!
Call: Are we going to beat this team?
Response: You bet!

If time permits, have volunteers teach their verses to the class. If they wish, students can add percussion parts, clapping, or choreography.

CLOSE

1 To end the lesson, gather classroom drums and ask students to play the following simple rhythm patterns along with "Thina Simnqobile" from 1:24–2:23. Start a small group with steady hand claps as shown below and then introduce the low conga part first. Then add the medium conga if students can perform the first two parts accurately.

2 Give all students the opportunity to play the different patterns. Once students feel comfortable, allow some to improvise rhythms during the section of the song when the clapping occurs on every beat (1:24–1:40 and 2:11–2:23).

Student Worksheet: Sounds of South Africa

Name_______________ Grade____ Music Teacher____________

Call and Response

The song "Thina Simnqobile" is in the *isicathimiya* (iss-ee-kah-tah-MEE-yah) style, which is similar to the call-and-response form that you may know from singing African-American spirituals or other songs. In the call-and-response form, a soloist sings the call and a chorus responds.

See if you can create your own call-and-response verses below. Keep in mind that the verses repeated by the chorus (the Response) should be simple enough for a group to sing or say together easily. The Response verse can even repeat the Call verse, just as in "Thina Simnqobile."

Use one of the suggested environmental settings below to help you think of ideas for your call-and-response verses, or you can compose verses for a setting of your choice.

School pep rally **Classroom** **Baseball game/other sporting event**

Holiday celebration **Birthday party** **Family picnic** **School bus**

Call: ____________________ Response: ____________________

Call: ____________________ Response: ____________________

Call: ____________________ Response: ____________________

Call: ____________________ Response: ____________________

Call: ____________________ Response: ____________________

If you like, you can notate the rhythm of your call-and-response verse on staff paper. If you are incorporating choreography, you can indicate the moves that accompany the verses on the back of this page.

Recommended Listening

South African Music

Pop/Mbaqanga/Kwela

Mahotella Queens ◎ ***Best of the Mahotella Queens Township Idols*** (Wrasse)

The Soul Brothers ◎ ***Jive Explosion*** (Earthworks)

Mahlathini ◎ ***The Lion of Soweto*** (Shanachie)

The Boyoyo Boys ◎ ***Back in Town*** (Rounder)

Johnny Clegg & Savuka ◎ ***Third World Child*** (Capitol)

Various Artists ◎ ***The Indestructible Beat of Soweto, Vol. 4*** (Earthworks)

Mbube/Isicathimiya

Ladysmith Black Mambazo ◎ ***The Best of Ladysmith Black Mambazo*** (Shanachie)

Ladysmith Black Mambazo ◎ ***Shaka Zulu*** (Warner Bros.)

Various Artists ◎ ***Mbube Roots: Zulu Choral Music from South Africa 1930s–1960s*** (Rounder)

Folk/Jazz

Miriam Makeba & the Skylarks ◎ ***The Best of Miriam Makeba*** (BMG International)

Hugh Masakela ◎ ***Grazing in the Grass: The Best of Hugh Masakela*** (Columbia)

Dudu Pukwana & Spear ◎ ***In the Townships*** (Earthworks)

Miriam Makeba ◎ ***The Click Song*** (Esperance)

Kwaito

Bongo Maffin ◎ ***Bongolution*** (Lightyear)

Various Artists ◎ ***Kwaito: South African Hip Hop*** (Stern's Africa)

Sounds of Mexico

H*ye-aye-eeee, ha ha ha ha!* If you've ever attended a Mexican festival or been to a Mexican restaurant with music, you've probably heard the unforgettable cry of the Mexican **mariachi** (mar-ee-AH-chee). With its combination of trumpets, violins, guitars, and other stringed instruments backing powerful singers, a mariachi performance is a unique musical experience. But mariachi is only one of Mexico's music styles. Read on to discover the many other genres our southern neighbor has to offer.

Proud Past

Mexicans are very proud of their history, and of their INDIGENOUS cultures in particular. The AMERINDIAN peoples who lived in Mexico before the arrival of the Europeans in the 1500s include the Mayans and the Aztecs. Originating around 1800 B.C. on the Yucatán Peninsula near where many Americans take vacations today, the Mayans created huge pyramids such as those at Chichén Itzá, and also invented a written language. They were followed by the Aztecs, whose civilization began around 1325. Yet because there is no evidence left, the music of both cultures is SHROUDED in mystery.

When the Spanish explorer Hernán Cortés landed near present-day Veracruz in 1519, everything changed for the Mexican people. The Spanish conquistadors set out to destroy nearly all examples of Mexican culture, including their music and musical instruments. Without their own instruments, native Mexicans were forced to adopt the stringed instruments of the Spanish, such as the violin and guitar. They later MODIFIED some of these instruments according to their own designs. The Mexicans also inherited their multipart harmonies in thirds and sixths from the Spanish.

El Rey: Mariachi

Today, the **mariachi** style reigns as *el rey* (the king) in Mexican music. Mariachi originated in the state of Jalisco (ha-LEES-ko) in western Mexico in the late 1800s. Popular musical groups featuring two violins, a small guitar called a ***vihuela*** (vee-WAY-lah), and a bass guitar called a ***guitarrón*** (gee-tar-RONE), started to be called "mariachis."

In 1934, Mexican president Lázaro Cárdenas invited the most famous mariachi group in Jalisco—Mariachi Vargas de Tecalitlán—to play at his inauguration. They were an instant success, and afterward mariachi music started to become popular throughout Mexico. You may wonder why mariachi performers wear such colorful costumes.

These come from the dress of the Mexican *charros*, or cowboys, who raised cattle in Jalisco.

There are two main roles for the instruments in a mariachi group: melody and rhythm. Instruments such as the trumpet and violin play the melody. The guitar, *vihuela,* and *guitarrón* are the rhythm instruments. The *vihuela* has five strings and is strummed with a long fingernail, which produces a crisp chordal sound. The six-string *guitarrón* keeps the beat—either straight or SYNCOPATED—and anchors the music with its low, deep sounds.

Guitarrón

Sing Us a *Son*

Many folk styles in Mexico have names that begin with *son,* which is a short way of saying "song style." So, ***son jarocho*** (ha-RO-cho) means "song style from the state of Veracruz" (*jarocho* is slang for things from Veracruz). You have heard a *son jarocho* song, but you probably didn't know it. In the United States, the late, great rock-and-roll musician Ritchie Valens took a *son jarocho* song and basically made it "rock." That song was "La Bamba."

Like all *son jarocho* tunes, the original version of "La Bamba" featured three instruments specific to the genre: the ***jarana*** (ha-RAH-na), a small guitar with 8 strings; the ***requinto*** (ray-KEEN-toe), a 4-string guitar that plays the melody; and the ***arpa,*** a 32- to 36-string harp that plays bass and melody. Wow, Mexican music features a lot of different guitars!

African Heritage

Down south in the states of Oaxaca and Chiapas, where active volcanoes still rumble from time to time, a very interesting style of music is played called ***son istmeño***. This style comes from West Africa, across the Atlantic Ocean. The key instrument in *son istmeño* is the **marimba**. When enslaved West Africans were brought to Mexico by the Europeans in the 1600s, they brought along the marimba, which is similar to the xylophone. *Son istmeño* songs can feature up to four marimbas.

Some people think that Mexicans invented the marimba, but one thing about the instrument seems to point to Africa: In both West Africa and southern Mexico, marimbas have a hollow tube below each wooden key that vibrates when the key is hit. No other marimbas in the world have this unique feature.

Strike Up the *Banda*

You may not associate Mexico with Germany, but German music had a big influence on one particular style of Mexican music: ***banda*** (BAHN-da). Today, *banda* music may be the most popular Mexican style. It's so popular, in fact, that you can hear *banda* music on Spanish radio stations throughout the United States.

What does it sound like? One clue is in the name. As in marching bands, *banda* musicians play brass instruments such as the trumpet, sousaphone, and trombone. And just like German "oom-pah" music, the sousaphone plays the low "oom" and the trumpets play the "pah-pah." These days, many *banda* songs are ***corridos,*** which are ballads that describe life in border towns. Though most *banda* songs are about "macho" topics, female singer Jenni Rivera proved that you don't have to be a macho man to sing *banda.*

The Germans also snuck something else into Mexican music: the accordion. In Mexico, accordion music is called ***norteño*** (nor-TEN-yo) and is almost exclusively for dancing. One of the most famous *norteño* groups is Los Tigres del Norte. Other popular Mexican music styles today are ***cumbia,*** another dance music style that originally came from Colombia in South America, and styles influenced by the United States, such as ***rock en Español*** (Spanish rock) and hip-hop.

South of the Border

So is mariachi just the beginning of Mexican music? *Sí, sí!* (Yes, yes!) From the deserts of the north to the high mountains of the Sierra Madre in the east to the rainforest jungles in the south, a different music style can be found in every region of this VIBRANT country. And it can all be heard just south of the border!

listen to: MARIACHI ("LOS ARRIEROS") AND SON JAROCHO ("LA GUANABANA")

Sounds of Mexico

Objectives

- Practice keeping a steady beat with different meters (NS 2)
- Learn about music from Mexico (NS 9)
- Identify shifting meter in a *son jarocho* and a mariachi song (NS 6)
- Create a speech piece with shifting meter (NS 4)

Materials

- *Music from Around the World* CD
- Student Worksheet: Music Review Journal (page 16)

START

1 To warm up the class, play a game of "Pass the Beat." This game can help students develop a steady beat, listening skills, and group dynamics while introducing ***sesquialtera,*** or shifting meter, a feature of many Mexican music styles. Have students stand in a circle. Establish a tempo by patting four beats on your legs. On the fifth beat, clap and turn to the person next to you. He or she must clap the next (6th) beat in tempo ("receiving" the beat from you), turn, and clap again facing the next person ("passing" the beat to him or her).

Experiment with various tempos and then select different students to set the tempo and direction of the claps.

2 Explain to students that many songs have meters that are duple or triple. This means the number of beats in a measure can be divided into two groups (duple) or three groups (triple). Lead the class in counting and clapping 3/4 (triple), and 4/4 (duple) meter, as well as 6/8 (duple; two groups of three) meter.

Share that some styles of music, like mariachi music from Mexico, switch back and forth between duple and triple meters. In Mexico, this is called *sesquialtera*. Have students try this while passing the beat, counting "ONE-two-ONE-two-ONE-two-three-ONE-two-three-ONE-two-ONE-two," etc., and clapping on each downbeat (ONE).

DEVELOP

1 Have students read "Sounds of Mexico" either silently, softly aloud in pairs, or taking turns out loud, depending on their abilities.

LITLINK Review the highlighted vocabulary words in the article. Ask for volunteers to try to determine the meaning of each word from its context. **[This reinforces vocabulary.]**

INDIGENOUS—original inhabitants
AMERINDIAN—inhabitants of Mexico before colonization
SHROUDED—hidden
MODIFIED—changed, altered
SYNCOPATED—a rhythm that stresses the off-beat in the meter
VIBRANT—pulsating with life or energy

2 Tell students that they are going to hear a *son jarocho* song called "La Guanabana" performed by Siquisirí from Veracruz, Mexico. Share that this all-string group plays two *requintos*, two *jaranas*, and an upright bass in place of the more traditional harp.

Explain that this piece features a slightly different *sesquialtera* from the one they learned at the beginning of class. Write the following pattern on the board. Ask students to clap on each "1" beat. (**Note:** There should be no pause between the first and second lines.)

1 2 3 **1** 2 3 **1** 2 3 **1** 2 3

1 2 **1** 2 **1** 2 **1** 2 **1** 2 **1** 2

3. Ask students to identify which line is triple and which is duple meter. (The first line is duple because there are two groups of three. The second is triple because there are three groups of two.)

4. Play "La Guanabana" (CD track 3) and have students listen for the above pattern, which occurs rapidly in the beginning of the song (0:10–0:41). (The pattern can be identified most clearly in the bass part, which stresses the "1" beats.) Then lead them in chanting the pattern and clapping along.

5. Now tell students they are going to hear a piece of mariachi music called "Los Arrieros," performed by a group from Los Angeles called Mariachi Los Camperos. Distribute copies of the Music Review Journal worksheet. Review any music vocabulary that might be unclear (monophonic, polyphonic, etc.). Play "Los Arrieros" (CD track 4) and have students complete the worksheet as they listen (see "Worksheet Answers" at right). Discuss their impressions of the piece as a class.

6. Review the music concept of *sesquialtera.*

ASK **Do you think "Los Arrieros" uses sesquialtera?** (Yes)
How can you discover if it does? (Listen for sections that have duple meter and other sections that have triple meter.)

Replay the following three excerpts from the song. Ask students to identify whether the meter of each one is duple or triple. Suggest that they pat the beat softly in their laps as they repeatedly count "1-2" or "1-2-3" and see which fits.

0:16–0:24—triple (3/4)
0:55–1:03—duple (4/4)
1:35–1:43—triple (3/4)

CLOSE

1. Write the phrase below on the board. Have the class echo you as you clap and speak the rhythm.

ASK **Which measure uses duple meter and which uses triple?** (The first uses duple because there are *two* groups of three; the second uses triple because there are *three* groups of two.)

If students have trouble identifying triple and duple meter, count the rhythm as 1-2-3, 2-2-3, 3-2-3, 1-2, 2-2, 3-2, etc.

2. Divide students into small groups and have each group compose a short verse piece like the one above made up of 2 phrases comprising 12 beats each. They should test whether a duple or triple meter works better with certain words.

3. Invite each group to perform their speech piece, clapping the duple and triple rhythm measures. If time allows, ask volunteers to teach the class their piece.

Worksheet Answers
Style—mariachi; instruments—violin, *guitarrón,* guitar, *vihuela,* trumpet; melody—arched; meter—(varies); beat—changing; rhythm—syncopated; texture—polyphonic

Student Worksheet: Mexico

Name____________ Grade____ Music Teacher__________

Music Review Journal

This music is an example of which style?

____

____

My first response to this recording was:

____

____

After talking about the music and listening again, my response was:

____

____

What I Heard

The instruments played include (name as many as you can):

____

____

____

The vocal/instrumental performances are (circle one):

◎ boring ◎ good ◎ interesting ◎ exciting

This is especially true of the __________ part, which sounds like ______________

I would describe the melody as:

____

The meter of this music has ______ beats.

The beat is (circle one):

◎ steady ◎ changing

The rhythm is (circle all that apply):

◎ even ◎ uneven ◎ syncopated

The texture of the music is (circle one):

◎ monophonic ◎ homophonic ◎ polyphonic

The ________ mood of this music is ______ created by:

____

What makes this music different or special? ______________________

What I Think

My rating for this recording (from 1 to 10, 10 being the highest) is ____.

I gave it this rating because: (explain fully and clearly based on your answers above)

____

____

____

____

____

____

____

____

Recommended Listening

Mexican Music

Mariachi/*Son Jarocho*

Mariachi Vargas de Tecalitlán ◎ ***The Best of Mariachi Vargas de Tecalitlán*** (Sony International)

Jorge Negrete ◎ ***15 Exitos Inmortales*** (Sony International)

Lola Beltrán ◎ ***10 Años: Un Recuerdo Permanente*** (Warner Music Latina)

Miguel Aceves Mejía ◎ ***Lo Mejor de lo Mejor*** (Sony International)

Graciana Silva ◎ ***Sones Jarochos*** (Corason)

Traditional Music From Mexico ◎ ***Son Jarocho de Tlacotalpán*** (ARC Music)

Ritchie Valens ◎ ***The Best of Ritchie Valens*** (Del-Fi)

Norteño

Los Tigres del Norte ◎ ***Herencia Musical: 20 Norteñas Famosas*** (Fonovisa)

Los Huracanes del Norte ◎ ***Corridos*** (Fonovisa)

Los Rieleros del Norte ◎ ***En Concierto*** (Fonovisa)

Banda

Banda el Recodo ◎ ***En Vivo*** (Fonovisa)

Banda Machos ◎ ***16 Reales Hits*** (Warner Music Latina)

Jenni Rivera ◎ ***Parrandera, Rebelde y Atrevida*** (Fonovisa)

Cumbia

Fito Olivares ◎ ***25 Aniversario: Famosas y Sabrosas*** (Univision)

Selena ◎ ***Amor Prohibido*** (EMI Latin)

Kumbia Kings ◎ ***Greatest Hits "Album Versions"*** (EMI Latin)

Sounds of Brazil

Here's a quick pop quiz: What is the only country in South America where the people speak Portuguese? Where is the world's largest tropical rainforest? Where does samba come from? The answer to each question is…Brazil!

Brazil's Beginnings

Brazil is the largest country in South America; it takes up almost half of the total land area of the continent. Hot, dense rain forests cover approximately half the country, and the largest river in the world, the Amazon, flows through the northern region. São Paulo, Brazil's largest city, is home to more than 10 million people—that's more than in New York City!

The history of Brazil had an important role in shaping the country's music. The first people to settle there were the Native Americans, who arrived more than 10,000 years ago. Their descendants still live in parts of the Amazon River basin, and their influence can still be seen in some Brazilian folk dances and instruments. In 1500 the Portuguese arrived and began to COLONIZE the country. They brought the Portuguese language that nearly everyone in Brazil now speaks—as well as their melodies, song lyrics, and stringed instruments.

Because they needed workers to CULTIVATE the land, the Portguese starting bringing slaves over from Africa in the mid-1500s. From then until 1888, when the slave trade was ABOLISHED, more than 3 million Africans were brought to Brazil, and with them came their musical HERITAGE. The syncopated rhythms, call-and-response song forms, and many of the percussion instruments heard in Brazil today are a direct result of the influence of these early African musicians.

Sounds of Samba

The eventual INTEGRATION of blacks, whites, and Native Americans in Brazil has resulted in an extremely diverse, open, and unique culture. **Samba** is a good example of this Brazilian cultural blend, and no music says "Brazil" more than this style. Today, samba is the living soundtrack of Carnaval (see sidebar, "Carnaval!").

The origins of samba—and all modern Brazilian music—began with the end of slavery. In the late nineteenth century, former slaves from the ***bairros*** (ghettoes) and ***favelas*** (shantytowns) of Rio de Janeiro formed ensembles made up of various drums and percussion instruments and started adding the melodies and harmonies of Portuguese music to the African rhythms This music eventually came to be known as samba, and it soon caught on with all types of people in Brazil.

Like rock or hip-hop, samba comes in many different

varieties, but features one common element—a steady "one-two-one-two" beat. You can hear this beat in Carnaval's ***samba-enredo*** (en-HAY-doo), where ensembles made up of hundreds of musicians parade through the streets. The beat can also be heard in the more melody-driven ***samba-canção*** (cahn-SAH-oh), where a singer is accompanied by a guitar and a ***cavaquinho*** (kah-vah-KEEN-yoh), which is like a ukelele. More recently, a new type of music called **samba-reggae** has appeared. Olodum, a group from Salvador da Bahia in the northeast, invented the style, which mixes samba with the rhythms of Jamaican reggae.

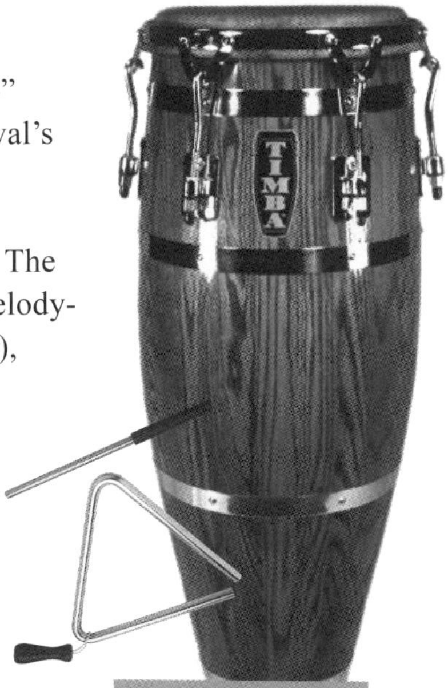

Timba **drum and triangle**

Bossa Nova and Other Beats

In the 1940s and '50s, young musicians from wealthy, beachside Rio de Janeiro neighborhoods like Ipanema started listening to the SPARSE, laid-back sounds of American cool jazz and fell in love. Musicians such as singer/guitarist João Gilberto, pianist Antônio Carlos Jobim, and lyricist Vinícius de Moraes began to adapt the sweet melodies and syncopation of samba to the style of cool jazz. And so, **bossa nova** (or "new beat" in Portuguese) was born. Beautiful songs like "Desafinado (Out of Tune)" and "Garota de Ipanema (The Girl from Ipanema)" became standards in both Brazil and the United States. Originally considered music for the upper class, bossa nova soon spread throughout the mass population of Brazil and the entire world.

Another sound often heard at the beach is the music of a dynamic performance style called ***capoeira*** (cah-PWEH-rah). Incorporating dance, martial arts, and music, *capoeira* presents two opponents who twist, spin, and cartwheel around each other to the music of a ***berimbau*** (BER-im-bow), a musical bow from Africa. The graceful moves of the dancers MELD perfectly with the repetitive, metallic pulse of the *berimbau*.

Another important style of music and dance from the northeast is called ***forró*** (foh-roh). A typical *forró* trio consists of a triangle, bass drum, and accordion.

MPB and the Future

MPB, or ***música popular Brasileira*** ("Brazilian popular music"), is a term Brazilians use to ENCOMPASS all of the current popular music in Brazil. MPB includes many Brazilian as well as foreign styles, such as jazz, rock, funk, and R&B. Yet each region in Brazil has its own flavor of MPB, which helps keep Brazilian music fresh and exciting.

MPB began in the 1960s, when the military government tried to suppress freedom of expression. Chico Buarque, Milton Nascimento, and the incredible female vocalist Elis Regina wrote and performed protest songs that added a modern twist to traditional Brazilian music.

Meanwhile, ***tropicália*** artists including Caetano Veloso, Gilberto Gil, and Tom Zé combined poetic political lyrics with rock-and-roll, jazz, folk, samba, and bossa nova. MPB's popularity has helped Brazil's music industry to become one of the biggest in the world.

The music of Brazil is so rich, it's nearly impossible to cover everything. But with just one listen, it's easy to get the sound of samba into your soul, the beat of Bahia under your feet, and the melodies of MPB planted in your mind! ◎

Carnaval!

Each year during the four days before Ash Wednesday in February or March, the entire country of Brazil takes a vacation. Instead of resting, though, everyone parties like there's no tomorrow! It's Carnaval, and the streets are filled with people, parades, and music. Gigantic, colorful floats and samba ensembles wind through the streets playing instruments like ***surdo*** bass drums, ***caixa*** snare drums, long ***timba*** drums, whistles, tambourines, and more. It's an unforgettable sound and spectacle! If you're ever lucky enough to attend, you'll hear the music of samba superstars like **Martinho da Vila** and **Paulinho da Viola** as well as groups like **Olodum** and **Banda Reflexu** playing throughout the celebration.

◎ **listen to:** SAMBA ("PESO NO BALANÇA") AND BOSSA NOVA ("UMA PRECE")

Sounds of Brazil

Objectives

- Learn about the music styles, instruments, and history of Brazil (NS 9)
- Identify two different styles of Brazilian music (NS 6)
- Perform a samba rhythm (NS 2)
- Perform and identify 4-note jazz chords (NS 2, 6)

Materials

- *Music from Around the World* CD
- Student Worksheet: Compare & Contrast (see page 22)
- Conga or bass drums, piano or mallet instruments

START

1. Divide the class into pairs or small groups. Have each group create a list of three American styles of music that are very different from one another, such as classical, country, and rap. They should be able to describe the differences among the three styles using at least two musical elements—for example, instrumentation, rhythm, lyrics, vocal style, etc. Have students share their lists with the rest of the class.

   ASK **What makes the styles different? How do you think styles so different could be from the same country?** (Different cultures, different tastes, different age groups, historical reasons, etc.)

2. Distribute the Compare & Contrast student worksheet. Tell students they are now going to listen to and compare two Brazilian pieces: Luiz Bonfá's "Uma Prece," and Wilson Moreira's "Peso No Balança." Share that Brazil is the largest country in South America and, just like the United States, has many different music styles. Play 20-second excerpts of each piece (CD tracks 5 and 6) and give students time to fill out the worksheet. Discuss the differences between the two pieces as a class.

DEVELOP

1. Have students read "Sounds of Brazil" on pages 18–19 either silently, quietly aloud in pairs, or taking turns out loud, depending on their abilities.

   LITLINK Review the highlighted vocabulary words. Ask for volunteers to determine the meaning of each word from its context. **[This reinforces vocabulary.]**

   COLONIZE—establish an area and community with common rules
   CULTIVATE—prepare for the raising of crops
   ABOLISHED—ended the practice of
   HERITAGE—family or cultural traditions passed down
   INTEGRATION—blending into the larger whole
   SPARSE—having very few elements
   MELD—combine

2. Tell the class that the two styles of music they heard earlier were samba ("Peso No Balança") and bossa nova ("Uma Prece"). Remind students that samba is built upon a steady "one-two-one-two" rhythm pattern that is played on *surdo* bass drums tuned to different low pitches. Point out that two basic bass pitches are played in the rhythm: one low note (the

marcacao reposta) is played on beat 1 by the *surdo* drum, and an even lower-pitched note (the *marcacao primeira*) is played on beat 2.

3. Introduce the following rhythm part to your students to play on either bass or conga drums, if available. They can also use soft mallets or their hands on a desk or table. Try to find two drums that have different pitches. (The higher-pitched drum is written on the C line; the lower-pitched drum is written on the A line.)

4. Play "Peso No Balança" (CD track 6) again and ask students to listen for and then play the *surdo* rhythm along with the recording.

5. Challenge students to recall the historical connection between samba and bossa nova they read about in the article. (Bossa nova composers tried to mix the sounds of jazz with the rhythm of samba.)

6. Share with students that one of the distinguishing characteristics of jazz is the use of chords that have more than three pitches. In American and Brazilian folk music, most chords use three pitches: root, third, and fifth. Explain that jazz often adds additional pitches, like the seventh, ninth, or even thirteenth notes. These numbers identify the interval, or number of pitches, above the root.

7. Play the following notes for students, alternating between an arpeggio and a block chord:

   C E G (C major triad)
   C E G B (C major 7 chord)

   ASK How would you describe the difference between the triad and the 4-note chord? (The 4-note chord sounds more full, less resolved, more emotional, etc.)

   Explain that the second chord is called C major 7 because the fourth note (B) is the interval of a seventh above the root (C).

8. Give students an opportunity to create different four-note chords. Ask volunteers to play the basic three-note triad and then add different notes above, in between, or below the triad on piano or using mallet instruments.

   Ask students to compare the sounds of the different four-note chords they create. Challenge them to name the interval between the root and the fourth note of the chord.

9. Now play the intro to "Uma Prece" (CD track 5) again and ask students to listen for the first chord of the song (G♯–D–G–C). Then have them listen for the chords in the first part of the verse (0:09–0:22) (F major, A major, D minor).

   ASK Do you think the chord in the intro is a 3-note or 4-note chord? (4-note chord) **What happens after the intro?** (The chords switch to 3-note chords.)

CLOSE

1. Ask students to recall which other music style besides jazz contributed to bossa nova (samba). Have volunteers play the *surdo* rhythm they played earlier in class along with this bossa nova tune.

   ASK Did the *surdo* rhythm fit the "feel" of the bossa nova piece? Why or why not? (No; it felt too slow; the bass of the bossa nova piece didn't match the surdo rhythm; etc.)

Worksheet Answers

Selection A ("Uma Prece"): style—bossa nova; instrument—guitar; acoustic; minor key; slow tempo; 2/4 meter; changing beat; even rhythm; mood—pensive, etc.

Selection B ("Peso No Balança"): style—samba; instruments—guitars, drums, percussion, *cavaquinho*, accordion, clarinet, flute, sax, trombone, vocals; acoustic; major key; moderate tempo; 4/4 meter; steady beat; uneven rhythm; mood—dancing, etc.

Student Worksheet: Brazil

Name______________ Grade____ Music Teacher__________

Compare & Contrast

Music Selection A: _______________
By:_______________
Music style:_______________
Instruments played:_______________
____

Acoustic or electric? _______________

The key is: major / minor (circle one)

The tempo is:
____ fast
____ moderate
____ slow

The meter has ____ beats.

The beat is:
___ steady
___ changing

The rhythm is:
___ even
___ uneven
___ syncopated

Overall mood created by these elements is:
____
____
____

Music Selection B: _______________
By:_______________
Music style:_______________
Instruments played _______________
____

Acoustic or electric? _______________

The key is: major / minor (circle one)

The tempo is:
____ fast
____ moderate
____ slow

The meter has ____ beats.

The beat is:
___ steady
___ changing

The rhythm is:
___ even
___ uneven
___ syncopated

Overall mood created by these elements is:
____
____
____

Brazilian Music

Bossa Nova

João Gilberto ◎ ***Amoroso/Brazil*** (Warner Bros)

Antonio Carlos Jobim ◎ ***Wave*** (A&M)

Luiz Bonfá ◎ ***Solo in Rio 1959*** (Smithsonian Folkways)

Astrud Gilberto ◎ ***Astrud Gilberto's Finest Hour*** (Verve)

Samba

Jason Miles ◎ ***A Love Affair: The Music of Ivan Lins*** (Telarc)

Various Artists ◎ ***32 Sambas: Carnaval Do Brazil*** (Orfeon)

Various Artists ◎ ***Brazil Roots Samba*** (Rounder)

Forró & Capoeira

Various Artists ◎ ***Brazil Classics Vol. 3: Forró, Etc.*** (Luaka Bop)

Various Artists ◎ ***Afro-Brazilian Music in Bahia: Capoeira, Samba Candomblé*** (Berlin Phonogramm)

MPB

Milton Nascimento ◎ ***Lo Borges Clube Da Esquina*** (World Pacific)

Chico Buarque ◎ ***Construçao*** (Philips)

Tom Zé ◎ ***Brazil Classics Vol. 4: The Best of Tom Zé*** (Luaka Bop)

Gilberto Gil ◎ ***Gilberto Gil (Frevo Rasgado)*** (Mercury)

Caetano Veloso ◎ ***Caetano Veloso (A Little More Blue)*** (Philips)

Sounds of the Islands

Have you ever wanted to take a cruise through the Caribbean? Imagine lying back in a chaise on the deck of a huge cruise ship, enjoying the warm weather, blue water, and gorgeous sunsets. Along the way, your ship stops at beautiful islands where different styles of great music welcome you.

While we can't send you on a real cruise, we can take you on a virtual trip through the Caribbean, stopping at some very musically alive locales: Jamaica, Hispaniola, and the related isles of Trinidad and Tobago. Let's set sail!

Maracas

Roots of Reggae

Your first stop on our virtual cruise is Kingston, Jamaica, where you have two whole days to explore the city. You head straight for the streets of Trench Town, hometown of the famous Jamaican musician Bob Marley. In the 1970s, Marley, along with his group the Wailers, helped to make **reggae** one of the most popular styles of music in the world. Bob Marley's image, with his famous dreadlocks, can be found on T-shirts and posters worldwide, and his songs like "No Woman, No Cry," "Three Little Birds," and "Get Up, Stand Up," have become sing-alongs for people everywhere. Other famous reggae artists include Jimmy Cliff, Judy Mowatt, UB40, and Bob's son Ziggy Marley.

As you stroll down the narrow streets, you hear some reggae music pumping out of a CD store, so you stop in to check it out. Striking up a conversation with the friendly shop owner, you learn that the roots of reggae can be found in the Jamaican styles of *mento, ska,* and rocksteady. The Jamaican folk music ***mento*** is a mixture of African and European musical elements, which makes sense since the English colonized Jamaica and brought many Africans to the island.

In the early 1960s the style called ***ska*** became popular. It featured simple rhythms with an emphasis on the off beats (the 2 and the 4). *Ska* took *mento,* added instruments and sounds from American R&B, and sped everything up—way up. The most famous *ska* band was the Skatalites, whose rapid guitar riffs and funky horn parts inspire people even today. *Ska* went through a "second wave" in the late '70s, with bands such as Madness, the English Beat, and the Specials playing "two-tone *ska,*" named for its multiracial band lineups and black-and-white checkerboard-style clothing. The "third wave" of *ska* in the '90s brought bands such as Smash Mouth, Less Than Jake, and No Doubt to international acclaim.

Because it was so fast, *ska* was not easy to sing to, so

it was slowed down to create a more moderate version called **rocksteady**. That's when Desmond Dekker, the king of *ska,* helped put Jamaican music on the map with his worldwide hits "007 Shantytown" and "Israelites."

The emergence of *ska* and rocksteady led to the birth of reggae around 1968. As in its parent styles, reggae's offbeat chords always fall on beats 2 and 4. These chords are usually played on a guitar, though sometimes on a piano. In reggae, however, a heavy bass beat always "drops" on the third beat. So the rhythm pattern is: rest, chord, drop, chord. This simple reggae formula has been copied all over the world.

The store owner tells you that the reggae you are probably most familiar with is called **dancehall**. Dancehall is like reggae's HYPER cousin. It has many of the same rhythms as reggae, but the instruments used are usually electronic, such as keyboards, samplers, or synthesizers. Producers create ***riddims,*** which are repeating rhythms that can be used by many different artists. Some dancehall artists sing over the *riddims,* while others rap. The biggest stars of dancehall—including Shaggy, Sean Paul, Buju Banton, and Shabba Ranks—get people out on the dance floor all over Kingston.

The Call of *Compas*

After two wonderful days in Kingston, your ship sets a northeast course for the island of Hispaniola, which takes its name from the Spanish words *La Isla Española* ("the Spanish island"). This was the name Christopher Columbus gave the isle when he landed there in 1492. The island actually COMPRISES two countries—Haiti and the Dominican Republic. In Haiti, people speak French or Creole, which is a mix of French and African languages. In the Dominican Republic, Spanish is mainly spoken. This is because France colonized Haiti, while the Dominican Republic was colonized by Spain.

Although Hispaniola's two countries might seem very different today, their histories are INTERTWINED. Hispaniola's original inhabitants, Native Americans called the Arawak, no longer live on the island. But their musical LEGACY continues with the ***guïra,*** a small, cylindrical metal instrument. Rubbing a stick up and down the *guira's* ribbed sides makes the unmistakable "scrip-a-scrap" sound heard in music all around the Caribbean.

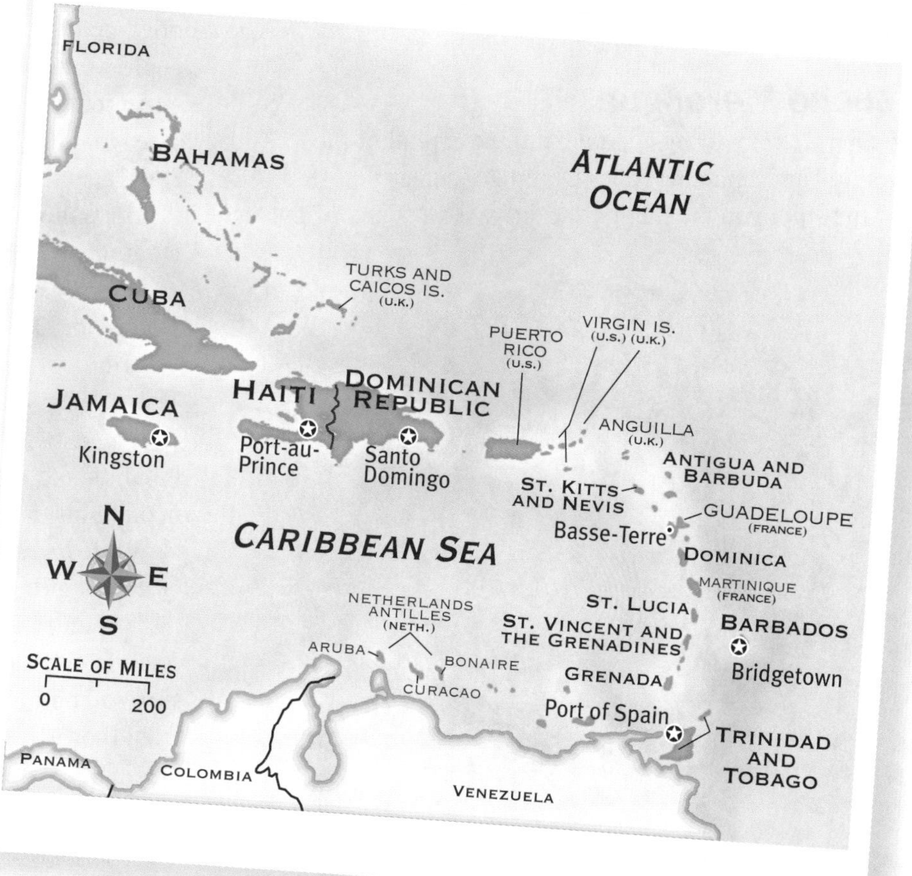

Your first port of call on Hispaniola is Port-au-Prince, the capital of Haiti. The lively streets of this waterfront city are filled with the heavy beats and slick guitar sound of popular radio bands like the System Band and Mizik Mizik. These groups play ***compas*** (comb-pah), the main style of Haitian popular music. *Compas* instrumentation was pared down from full big band to a smaller jazz combo, and can include electric guitar, bass, alto sax, drum and

(continued on next page)

("Sounds of the Islands" continued)
cowbell percussion, and sometimes accordion.

With waves of Haitian immigrants moving north, New York City became home to the most popular *compas* bands. These groups, such as Top Vice and Zin, will sometimes rap over a *compas* beat. Other popular styles in Haiti include ***comparagga,*** a mix of Jamaican reggae and *compas,* and ***twoubadou,*** a style similar to R&B in the United States.

Mucho Merengue

Your ITINERARY brings you next to the capital of the Dominican Republic, Santo Domingo, and as you step off the ship, you are met by the SIGNATURE sound of the country: ***merengue!*** *Merengue* is to Dominicans what rock is to Americans. A steady, even, four-beat drum pattern keeps the rhythm moving, while the vocals interact with the syncopated streams of the horn section. A *merengue* band can include 10 or more musicians playing trumpets, saxophones, bass, piano, accordion, **conga** drums, a two-sided drum called a ***tambora,*** maracas, and the *güira.* All of these ingredients are boiled up together in a steamy rhythmic soup that makes people want to do one thing: dance!

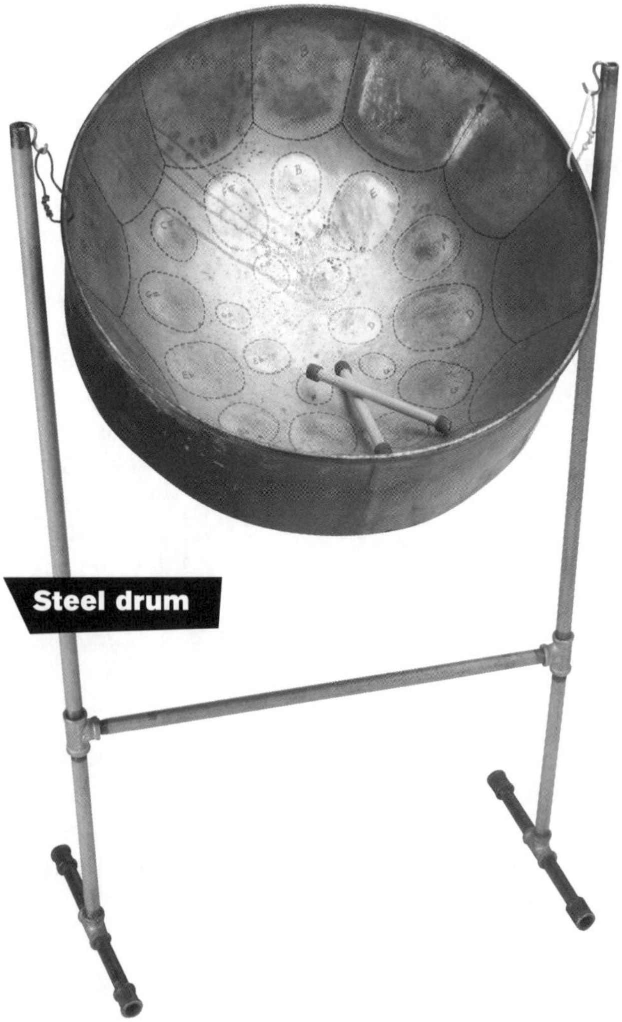

Steel drum

The roots of *merengue* music are believed to have come from the D.R.'s Hispaniola neighbor, Haiti. In the 1800s, a style of music called *méringue* was popular in Haiti and probably migrated across the island to the Dominican Republic.

Johnny Ventura helped to make *merengue* popular worldwide. Juan Luis Guerra, Fernando Villalona, and Wilfrido Vargas are other leading *merengue* artists. These days, young musicians in the D.R. are taking *merengue* in new directions. ***Merenhouse*** artists like Fulanito and ***merenrap*** lyricists like Proyecto Uno are making just a couple of the many new sounds.

While *merengue* and its offshoots represent their culture's glitzy pop music, Dominicans also love their country music, which is called ***bachata***. Like American country and blues artists, *bachata* musicians sing sad and lonely songs about lost love, losing a friend, or just plain bad luck. *Bachata* songs were originally a simple vocal set against a strummed guitar. Nowadays, *bachata's* MELANCHOLY stories are usually backed by an electric band.

Carefree Calypso

The last stop on your cruise takes you to one of the most musically interesting areas of the Caribbean: Trinidad and Tobago. This pair of islands is the birthplace of **calypso,** perhaps the most well known of all Caribbean styles. Calypso is a perfect "feel-good" blend—it's neither too fast nor too slow, and has pretty melodies, a pleasant mix of TIMBRES, and carefree lyrics. Lord Kitchener is one of the fathers of classic calypso. Mighty Sparrow follows in his wake, and Calypso Rose is one of the female calypso artists making big waves.

Calypso originally featured just a singer and a guitar. Now, the trademark sound is the **steel drums,** called ***pan*** in Trinidad. After World War II, many empty oil barrels were left on local beaches. Musicians discovered that you could play different pitches by hitting the tops of the barrels, and that with a little shaping, you could play entire melodies on them. Today, the plingy metallic "raindrop" sounds of steel drums can be heard all over Trinidad and the other islands of the Caribbean.

In the 1970s, calypso began to fade with the rise of Jamaican reggae. But Trinidadian musician Garfield Blackman (aka "Ras Shorty I") decided it was time to bring calypso back. So he added Indian instruments such as the *sitar* and *tabla* and sped up the tempo for an energized new flavor. Thus was born ***soca***—the "(so)ul of (ca)lypso." With its heavy bass-drum beat and slippery cymbals hissing on top, *soca* is another Caribbean style made for dancing. As in reggae, the bass lays down a funky rhythm at the bottom of the mix.

Soca's biggest hit was Arrow's "Hot! Hot! Hot!" in 1983. Today, *soca* stars like Alison Hinds, Explainer, and Kevin Little are taking this music in interesting directions. ***Soca chutney*** is the sound of the Indian population in Trinidad—can you guess how it got that name? (Chutney is a popular CONDIMENT from India.) When people from India moved here in the 1800s, they wove *soca* together with the musical traditions of their ancestral home.

Another local style called ***rapso*** started in the 1970s, about the same time that rap EMERGED in the U.S. The founder of the genre, Lancelot Layne, mixed the rhythms of *soca* with rhyming poetry. Today, *rapso* is topping the charts, and groups like 3 Canal, Ataklan, and Black Lyrics share stories of Trinidad street life with their DEFT rhymes.

Worldly Wonders

Do you know which two Indian musical instruments, introduced into Caribbean calypso music by Ras Shorty I, are pictured as shadows here?

(Need a hint? See pages 46–47.)

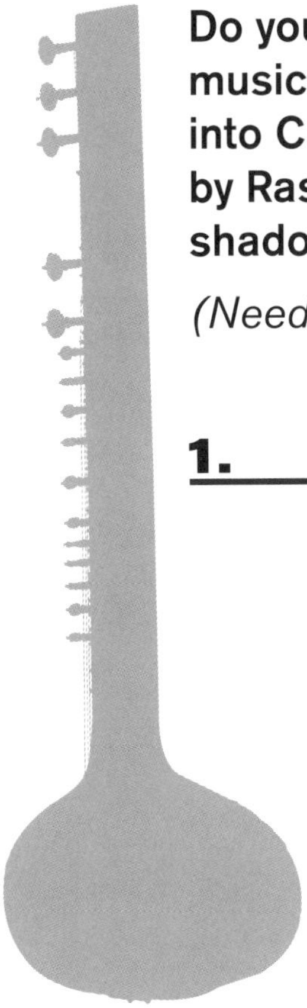

1. ____________________

2. ____________________

Transported By Music

Ahhhh, this musical cruise of a lifetime has to come to an end. Unfortunately, the ship can't stop at every island. But don't feel bad—now you have a lot more reasons to come back to the Caribbean!

listen to: REGGAE BY JUDY MOWATT, CALYPSO BY LORD KITCHENER, AND *SOCA* BY CALYPSO ROSE

Sounds of the Islands

Objectives

- Learn about the music of the Caribbean (NS 9)
- Identify musical elements of three Caribbean music styles (NS 6)
- Compose and perform a vocal-percussion pattern to accompany a Caribbean song (NS 1, 4)

Materials

- *Music from Around the World* CD
- Student Worksheets: *Compare & Contrast* (page 22), *Rhythm Building* (page 30)
- Unpitched instruments (bongos, wood blocks, claves, conga drum, etc.)

START

1 LITLINK Have students split up into pairs. Ask them to imagine being on an island in the Caribbean, and have each pair write down 10 adjectives that come to mind. (For example: warm, sunny, relaxing, blue, refreshing, etc.) As students create their lists, play "Let's Dance" by Judy Mowatt (CD track 7). Have each pair write three of their adjectives on the board. **[This reinforces vocabulary and fluency.]**

ASK **Why did you choose these adjectives? What type of music do you imagine you might hear in this region?** (Possible answers: reggae, dancehall, calypso, *merengue,* etc.)

Tell students that they were listening to a reggae song by Judy Mowatt, a singer from Jamaica. Share that she was one of the I-Threes, the background singers in Bob Marley's band the Wailers.

2 LITLINK Invite students to take turns reading "Sounds of the Islands" (pages 24–26) aloud. Review the highlighted vocabulary words. Ask for volunteers to use each word in an original sentence. **[This reinforces fluency and vocabulary.]**

HYPER—fast-paced, quick
COMPRISES–includes, is made up of
INTERTWINED—woven together
LEGACY—passing down something from past to present
ITINERARY—outline of a trip or tour
SIGNATURE—(characteristic) typical, distinguishing
MELANCHOLY—sad
TIMBRES—sound qualities, tone colors
CONDIMENT—a seasoning for food
EMERGED—became visible, came on the scene
DEFT—skillful

DEVELOP

1 Tell students that they will now hear two styles from Trinidad & Tobago (*soca* and calypso). Remind them that the trademark sound of calypso is the steel drum, and that *soca* features a heavy bass drum beat and funky bass rhythm.

2 Distribute copies of the Compare & Contrast worksheet. Using the worksheet as a guide, students should try to identify the characteristics and style of the following two tracks. Play the first 1:30 of both "Ju Ju Warrior" by Calypso Rose (CD track 8) and "Miss Tourist" by Lord

Kitchener (CD track 9) and have students fill in the worksheet to the best of their abilities.

Play "Ju Ju Warrior" (CD track 8) again and have students determine the meter of the song.

ASK **Does the song feel like the beats are in groups of two, three, or four?** (Four)
What is the time signature for this beat grouping? (4/4)
How many times per measure does the kick drum play? (Four)

Explain that this rhythm pattern is often called "four on the floor." Share that this is a common rhythm pattern for dance music in the Caribbean. Play "Let's Dance" (CD track 7) again and have students listen for the reggae rhythm pattern: rest, chord, bass drop, chord.

ASK **Which instrument is playing the chord, guitar or piano?** (Piano)
Does "Let's Dance" use the four on the floor rhythm pattern? (No)

3 (CD 8) Play "Ju Ju Warrior" (CD track 8) again and have students listen for the unusual sound that is heard from 0:22–0:27. Can they guess what made this sound? Hint: It can be played without hands. (It is a voice.) Point out that sometimes musicians use their voices for making sounds other than singing.

ASK **What styles of music or musicians can you think of that use voices for making accompaniment sounds?** (Examples: beatbox in rap, scat in jazz, Bobby McFerrin)

CLOSE

1 Distribute copies of the Rhythm Building worksheet. Have students reform in their original pairs, and tell them that they are going to create a "spoken accompaniment" pattern for "Ju Ju Warrior." Explain that they will compose the rhythm of their pattern first.

Take a few minutes to review as a class the names and durations of each note or rest pictured in the Rhythm Tool Kit. Ask a student to recall the time signature of the song (4/4), and direct students to create their patterns in the appropriate section on the worksheet.

2 (CD 8) When students have completed their patterns, ask them to choose two or three adjectives from either the list on the board or the list they created at the beginning of class and incorporate these words into their spoken accompaniment.

Give students about 10 minutes to compose their spoken word pieces. Play "Ju Ju Warrior" (CD track 8) quietly during their composition time to allow students to practice at the correct tempo.

3 Encourage each pair to perform their spoken accompaniment pattern both with and without the music of "Ju Ju Warrior."

4 Distribute percussion instruments to a group of four students. Direct each student to take one of the four beats in 4/4 time. For example, bongos take Beat 1, wood block takes Beat 2, etc. Let this group practice "four on the floor" rhythm, playing in sequence.

Then let the other students perform their spoken patterns with the group playing 4/4 accompaniment. As class time permits, rotate so each student plays percussion and performs vocally.

Worksheet Answers
"Ju Ju Warrior"—soca; drums, bells, whistles, bass guitar, horns, synthesizer; electric; major; fast; 4; steady; even. "Miss Tourist"—calypso; drum set, bell, horns, guitar, piano, acoustic bass, steel drums; acoustic/electric, major; moderate; 4; steady; even

"Worldly Wonders" Answers
1. sitar; 2. tabla

Student Worksheet: Caribbean

Name_______________ Grade_____ Music Teacher_____________

Rhythm Building

Building rhythm patterns isn't hard—like any builder, you just need to know how to use your tools! Here's a chance to practice composing rhythms out of notes and rests.

Rhythm Tool Kit

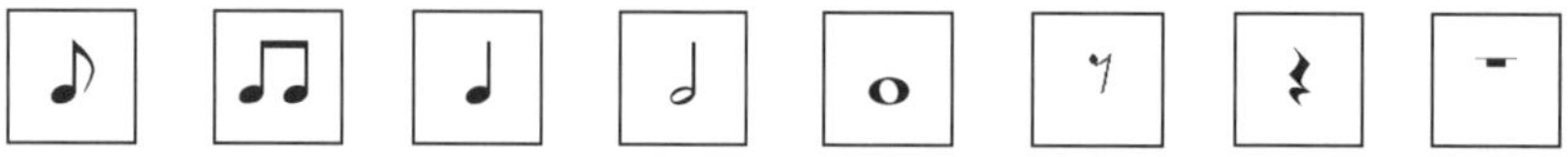

Rhythm in 4/4 time

A 4/4 time signature means there are 4 quarter-note beats per measure. Either a note or a rest can occur on each beat. Using the notes and rests in the rhythm tool kit above, create a simple 4-beat pattern. Here's a sample:

Now, you try:

You can play this rhythm pattern along with any music in 4/4 time!

Rhythm in 3/4 time

A 3/4 time signature means there are 3 quarter-note beats per measure. Here, too, either a note or rest can occur on each beat. Using the notes and rests in the rhythm tool kit above, create a simple 3-beat pattern. Here's a sample:

Now, you try:

You can play this rhythm pattern along with any music in 3/4 time!

Recommended Listening

Caribbean Music

Jamaica: Reggae/*Ska*

Bob Marley & The Wailers ◎ ***Legend*** (Tuff Gong)
Jimmy Cliff ◎ ***The Harder They Come*** (Mango)
Desmond Dekker ◎ ***Rockin' Steady*** (Rhino)
The Skatalites ◎ ***Foundation Ska*** (Heartbeat)
Shaggy ◎ ***Boombastic*** (Virgin)

Hispaniola: *Compas & Merengue*

Tabou Combo ◎ ***8ème Sacrement*** (Mini)
Various Artists ◎ ***Konbit: Burning Rhythms of Haiti*** (A&M)
Johnny Ventura ◎ ***40 Aniversario: En Vivo*** (WEA International)
Juan Luis Guerra ◎ ***Colección Romantica*** (Karen)
Wilfrido Vargas ◎ ***Raices*** (RCA International)
Various Artists ◎ ***The Rough Guide to Merengue & Bachata*** (World Music Network)
Fulanito ◎ ***Hombre Más Famoso de la Tierra*** (Cutting Records)

Trinidad & Tobago: Calypso & *Soca*

Lord Invader ◎ ***Calypso in NY*** (Smithsonian Folkways)
Mighty Sparrow ◎ ***Volume One*** (Ice Records)
Arrow ◎ ***Hot Soca Hot*** (Arrow)
Calypso Rose ◎ ***Soca Diva*** (Ice Records)
Black Stalin ◎ ***Roots Rock Soca*** (Rounder)
Shadow ◎ ***Columbus Lied*** (Shanachie)

WORLD MAP

Atlantic Ocean
THE CARIBBEAN
MEXICO
Pacific Ocean
BRAZIL
N
W
E
S
No

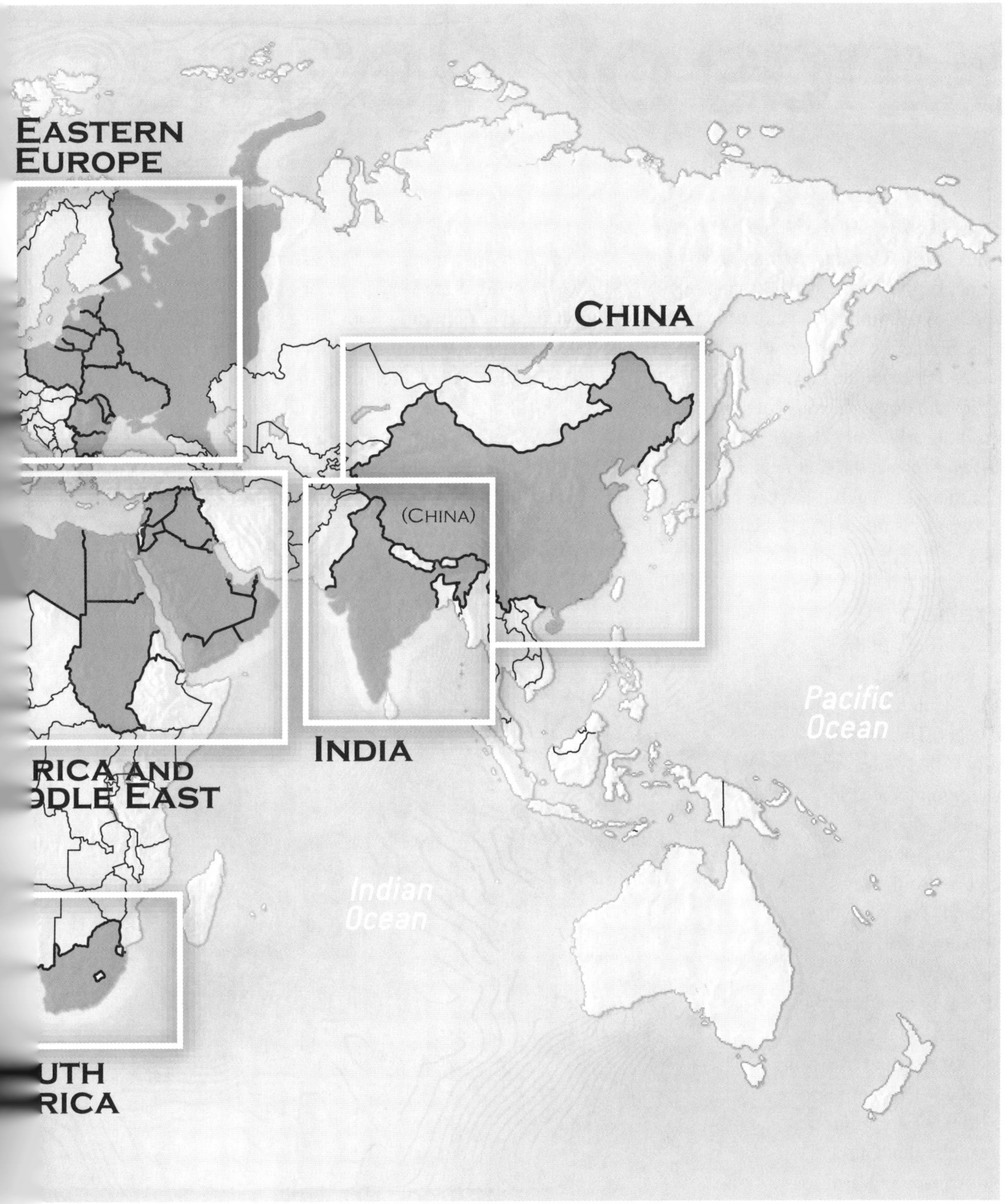
EASTERN
EUROPE
CHINA
(CHINA)
INDIA
RICA AND
DDLE EAST
UTH
RICA
Pacific
Ocean
Indian
Ocean

Sounds of Klezmer

The needle drops and the old record player comes to life with crackles and pops. A clarinet begins to play a slow, haunting melody. A mandolin soon comes in with choppy, short chords. Suddenly, the music picks up speed, and a bass, trumpet, and flute join in. The melodies move like a roller coaster up and down the musical scale, taking quick half-step turns at nearly every chance. What is this wild sound? It's ***klezmer*** music, the folk music of the Jewish people of Eastern Europe. Although it has been around for hundreds of years, *klezmer* is moving in directions that the old virtuosos never could have imagined.

Musical Migration

If you were to write a history of the Jewish people, you would need an incredibly big book—it stretches back thousands of years. The Jewish homeland, originally called Judea, was a dry, mountainous region located in the area now occupied by Israel and Palestine. In 72 A.D., the Romans who ruled Judea forced many of the Jews to leave their homeland. This resulted in the **diaspora,** the MIGRATION of the Jewish people across parts of Europe, Asia, and North Africa. Everywhere they went, they left their mark on the culture, including the music. One sign of this is the large number of great Jewish composers of classical music from Western Europe, including Felix Mendelssohn, Gustav Mahler, and Arnold Schoenberg. But for the story of *klezmer,* we must look to the Jews who settled in Eastern Europe.

Eastern European Jews shared a common language called Yiddish ("klezmer" comes from the Yiddish *klei zemer,* meaning "musical instrument"). Yiddish musicians, called *klezmorim*, were very popular and played for all kinds of occasions, including both Jewish and Christian weddings. The *klezmorim* absorbed musical influences from many different peoples, including the Turks and the Gypsies in Central and Eastern Europe and the Middle East. They often used an interval called an **augmented second,** in which there are three half steps between the second and third notes of the scale. In the key of C, this would be C, C♯, E, F, G. Play these notes, and you'll instantly hear an Eastern European sound.

Coming to America

From the late 1800s to the early 1900s, nearly three million Jews from Eastern Europe immigrated to the United States, many of them settling in New York City. Jewish musicians quickly began FUSING their traditional sounds with the styles they heard in their new home such as ragtime and the foxtrot dance rhythm. By a stroke of luck, they arrived just as the recording industry was getting started. Jewish musicians made hundreds of records during this period, and soon *klezmer* music could be heard everywhere.

In the 1800s, a *klezmer* group typically played only two instruments: the violin and the ***tsimbl***. The *tsimbl*—sometimes called a ***cymbalom*** or hammered dulcimer—looks like the inside of a piano. To play it, a musician uses two small mallets to strike the strings that stretch across a wooden soundboard. Later, the more standard *klezmer* lineup developed, which included a clarinet, upright bass, and two fiddles. With its reedy, mellow tone, the clarinet eventually became the signature sound of *klezmer* bands, especially in America. Two old masters of the American *klezmer* clarinet were Naftule Brandwein and Dave Tarras.

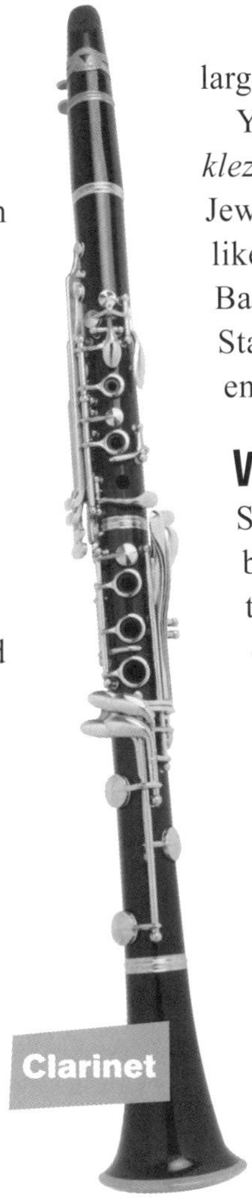

As with violinists, it is easy for a clarinet player to add emotion to a melody by bending the pitch. These NUANCES, which can express either humor or sadness depending on the player's emphasis, give *klezmer* its flavor. In the 1920s and 1930s, other instruments like guitar, trumpet, and flute were added to *klezmer* groups. At that time, it seemed like nothing could stop the *klezmer* expansion.

Cultural Crisis

Despite *klezmer*'s early popularity, by the 1940s many Jewish musicians had stopped playing this old-time style. They wanted to be modern, and *klezmer* music seemed out of fashion. Also, the music traditions of the European Jews had almost completely disappeared due to the Holocaust—one of the most terrible periods in human history. An estimated six million Jews were killed, largely destroying Jewish culture in most of Europe.

Yet through recordings and U.S.–based musicians, *klezmer* PREVAILED. By the 1970s, both Jewish and non-Jewish musicians began playing the style again. Groups like the Klezmorim and the Klezmer Conservatory Band, along with musicians like clarinetist Andy Statman and multi-instrumentalist Henry Sapoznik, encouraged a big *klezmer* comeback.

Where to Next?

Since the 1990s, *klezmer* in the United States has been about exploration. Its players honor the old tunes but also want to create new, cool music that expresses life today. This music for the new millennium can be heard from bands like Brave Old World and the Klezmatics, one of the most successful contemporary *klezmer* bands. Led by trumpeter Frank London, the Klezmatics have the chops to keep up with any musician, including one of the greatest classical violinists of all time—Itzhak Perlman. The legendary Perlman embraced *klezmer* in the 1990s and helped create the best-selling *klezmer* album of all time, *In the Fiddler's House*.

Today, every world music store or website has CDs of this continually evolving music. With sounds that are both MOURNFUL and joyous, ancient and cutting-edge, it's hard to tell what today's *klezmorim* have up their sleeves. Where will *klezmer* go to next? That's anybody's guess!

listen to: "KOLOMEYKE" AND "DYBBUK SHERS" BY THE KLEZMATICS

LESSON PLAN

Sounds of Klezmer

Objectives

- Review and identify a chromatic melody (NS 5, 6)
- Learn about *klezmer* music and the culture of Eastern European Jews (NS 9)
- Analyze the melody of two *klezmer* pieces (NS 6)
- Create a melody using the scale of a *klezmer* piece (NS 2)

Materials

- *Music from Around the World* CD
- Student Worksheet: *Pitch & Melody (p. 38)*
- Keyboard or mallet instruments

START

1. Write the two melodies below on the board and play them for your students.

ASK **What is different about the second melody?** (It uses sharps and flats.)

2. Explain that the second melody is chromatic. Chromatic melodies include notes that are not in the music's key, and are often indicated by the presence of sharp, flat, and natural signs.

ASK **What types of music do you know that often feature many chromatic notes? (**Jazz, blues, some classical music)

Share with students that they are going to learn about a style of Jewish music called *klezmer*. The melodies of klezmer often feature many chromatic notes.

3. Have students read "Sounds of Klezmer" (pp. 34–35) either silently or softly aloud in pairs, depending on their abilities.

LITLINK Check your students' understanding of the highlighted vocabulary words. Ask for volunteers to create new sentences using one of the words. **[This reinforces vocabulary.]**

MIGRATION—moving as a group from one area to another
FUSING—blending thoroughly; combining
NUANCES—small aspects that make something different
PREVAILED—overcame; won; survived
MOURNFUL—full of, or expressing, sadness

DEVELOP

1. Play "Kolomeyke" by the Klezmatics (CD track 10). Point out that, like many *klezmer* tunes, the form is made up of different sections of equal length. It also features two other typical *klezmer* characteristics: many chromatic notes and an uptempo melody.

LITLINK

ASK **Does the music sound like what you expected after reading the article? Why or why not?** (Answers will vary.) **[This reinforces fluency.]**

2 Distribute copies of the Pitch & Melody worksheet. Play 0:00–0:12 of "Kolomeyke" (CD track 10) and have students complete Part 1 of the worksheet for the melody of this section of the piece. Assist students in drawing the melodic contour by playing or singing the melody slowly—it is most clearly stated by the opening trumpet line—and having everyone trace the melody's shape in the air with their hand.

3 Play this section (0:00–0:12) of the melody again and give students time to work on Parts 2 and 3 of the worksheet. Allow them to use the piano or other pitched classroom instruments to help them answer the questions in Part 2. If your students know how to notate, challenge them to write the melody on the staff in Part 4 of the worksheet.

4 Now play "Dybbuk Shers," also by the Klezmatics (CD track 11), and have students describe the differences they hear between the two *klezmer* pieces. Encourage them to consider instrumentation, tempo, and style.

ASK **How would you describe the style of the violin in the beginning?** (Crying, wailing)

5 Review the meaning of the augmented second (an interval of three half-steps). Write the melody excerpt notated below on the board and ask for a volunteer to play it on piano or another classroom instrument. Have students identify the augmented second in the melody (it is between E♭ and F♯).

Play 2:06–2:24 of "Dybbuk Shers" (CD track 11) again and ask which instrument(s) they hear playing this melody (clarinet and trumpet).

CLOSE

1 Now write the following G minor scale and the *klezmer* scale in modified G minor, which is the scale used in "Kolomeyke," on the board:

2 Ask students to look at the two scales. Do they see any difference between them? (The modified G minor has an F♯).

3 Divide the class into pairs and have each pair work at keyboards or other pitched instruments to create their own melody pattern using the scale heard in the snippet above. Instruct them to compose four measures in 2/4 time. Allow each group 10–15 minutes to rehearse their melody. Ask volunteers to perform their *klezmer* melodies for the class.

Worksheet Answers

Part 1—This should be a simple line drawing that reflects the rising and falling of the melody. Drawings may vary.

Part 2—Starting note: D
ending note: D
key: G minor (0 sharps and 2 flats)
key sig:

Part 3—The melody is characterized by: steps, repeated notes, and short notes

Part 4—Here is the melody of 0:00—0:12 of "Kolomeyke":

Student Worksheet: Eastern Europe

Name______________ Grade____ Music Teacher____________

Pitch and Melody

1. Melodic Contours

In the space below, draw a line that shows the shape of the "chorus" or "main section" melody you hear. Let the line go up when you hear notes that get higher and let it go down when you hear descending notes.

2. Key Signature

Find the starting note and ending note of the melody you hear on a piano or keyboard.

The starting note of this melody is: ____
The ending note of this melody is: ____
This music is in the key of: ____
This key has ____ sharps and ____ flats

This key signature is indicated on a treble clef like this:

3. Melodic Qualities

This melody is characterized by (circle all that apply):

steps leaps repeated notes long notes short notes

4. Write the Melody!

Recommended Listening

Klezmer Music

The Klezmatics ◎ ***Possessed*** (Rounder)

The Klezmatics ◎ ***Rhythm & Jews*** (Flying Fish)

Itzak Perlman ◎ ***In the Fiddler's House*** (Angel)

Naftule Brandwein ◎ ***The King of the Klezmer Clarinet*** (Rounder)

Zev Feldman ◎ ***Jewish Klezmer Music*** (Shanachie)

Dave Tarras ◎ ***Yiddish-American Klezmer Music*** (Yazoo)

Klezmer Conservatory Band Live! ◎ ***The Thirteenth Anniversary Album*** (Rounder)

The Klezmorim ◎ ***East Side Wedding*** (Flying Fish)

Andy Statman Quartet ◎ ***Between Heaven & Earth: Music of the Jewish Mystics*** (Shanachie)

Flying Bulgar Klezmer Band ◎ ***Agada*** (Dorian)

Brave Old World ◎ ***Beyond the Pale*** (Rounder)

Golem ◎ ***Fresh Off Boat*** (Jdub)

Sounds of the Arab World

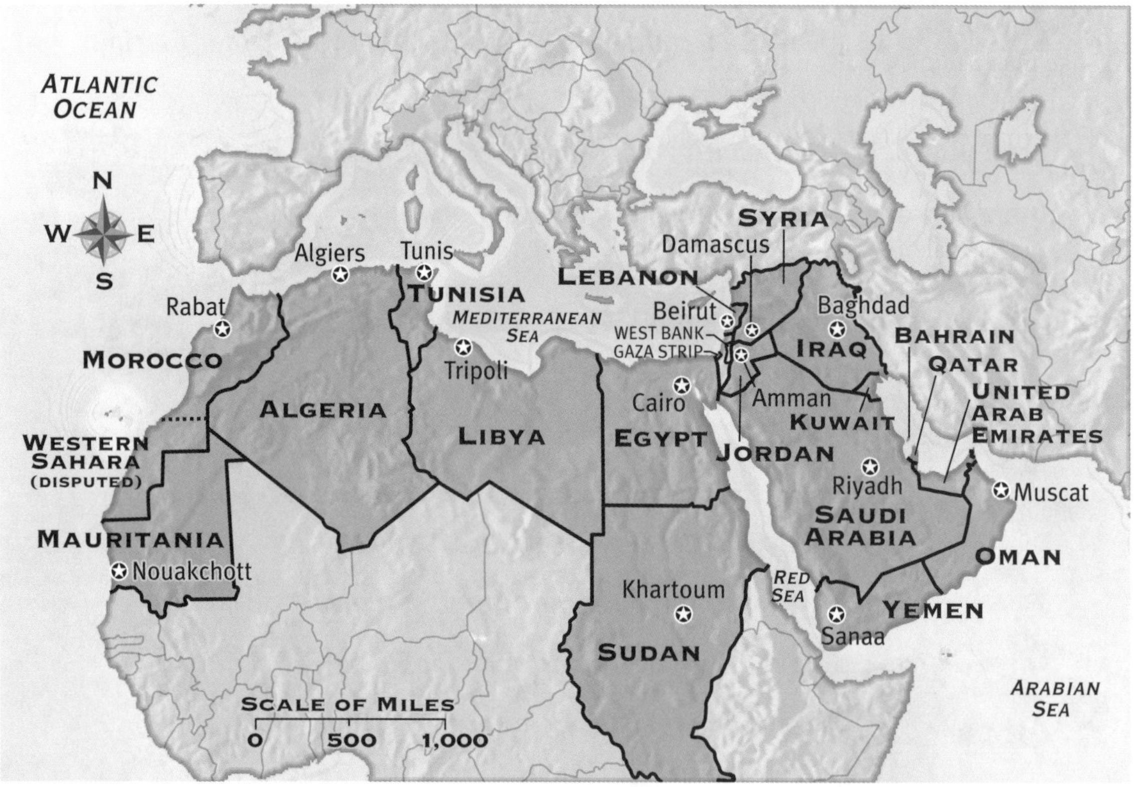

When you think of North Africa and the Middle East—the lands of the Arab people—what images come to mind? Deserts, camels, and turbans? ORNATE mosques and Muslims kneeling in prayer? Oil fields and fancy cars, or barren fields and poverty? What about the music? Do you hear strange melodies and wailing singers, and see belly dancers and unusual instruments?

To some extent, all of these ideas are accurate. But they are only a small part of the amazing diversity found among the Arab people. Along with their many races, religions, and professions, the Arab people also have a rich culture filled with wonderful art, food, and, best of all, music!

Unity and Variety

One thing that connects all Arabs is their language. Like English in the United States, the Arabic language has different DIALECTS, but people use a standard Arabic to communicate with each other. More than 250 million Arabs live in the area from North Africa to the Middle

East. This TERRITORY stretches from the Atlantic Ocean west of Africa all the way to Iraq in Asia, a distance of almost 3,000 miles (about the distance from New York to Los Angeles). The Asian part of the Arab world is called the ***Mashreq*** and the North African part (excluding Egypt and the Sudan) is known as the ***Maghreb***. As in the United States, the land varies from deserts to beautiful mountains and green fields.

Styles Like Ours

Like the music of other regions, Arabic music can be divided into folk, pop, and classical genres. Arab folk music varies depending on the part of the Arab world you are in. For example, the folk music in Sudan uses lots of drums and polyrhythms, like music further south in Africa. But up north in Lebanon, you might hear an accordion or a violin.

Pop music, though, is pretty similar throughout the Arab world, probably because everyone watches many of the same music videos. On Arabic television you can see videos of ***rai*** singers like male vocalist Amr Diab from Egypt or female vocalist Samira Said from Morocco, along with the latest music videos from the U.S.

On the classical side, female singers Umm Kulthum and Fairuz and singer/composer Mohamed Abdul Wahab are well-known all over the Arab world. You would recognize their ensembles: like here at home, Arab classical music features both large orchestras and small trios or quartets.

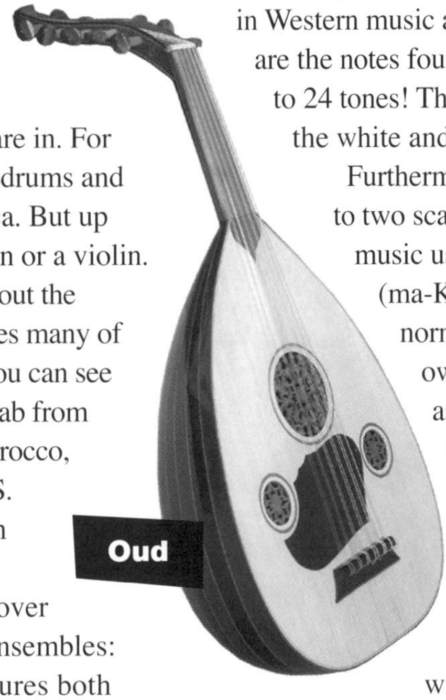
Oud

Interesting Instruments

In Arabic music, small ensembles called ***takhts*** (tockts) play both classical and folk, usually using the same combination of instruments. One of the most important of these is the Middle Eastern lute called the ***oud*** (ood). This fretless instrument has 10 to 12 strings which are plucked with a pick. The ***qanun*** (kuh-NOON) is another stringed instrument that looks like a small version of the inside of a piano; players use fingerpicks to pluck the strings. The ***ney*** (nay) is a reed flute that has a very breathy, airy sound; you play it by blowing across the top, like you would on a bottle. A violin or clarinet can also sometimes be heard in Arabic music.

Arabic music wouldn't have its excitement and distinctive sound without its unique percussion instruments. Two important ones are the ***riqq*** (reek) and the ***dumbek*** (DOOM-bek). The *riqq* looks like a tambourine, but the head is made out of fish skin! Players use complicated finger techniques to tap patterns on this skin. The *dumbek* is a very popular hand-held drum that is shaped like an hourglass. The hands of a *dumbek* player can sometimes move so fast that you can't see them, creating MESMERIZING rhythms.

Many Maqam

One reason that Arab melodies sound so different from ours is that they are made of more pitches. Most of the melodies in Western music are created from 12 tones. These 12 tones are the notes found in one octave. But Arab music uses up to 24 tones! These extra 12 notes would fall in between the white and black keys on a piano.

Furthermore, Western music usually sticks to two scales: minor and major. Arab classical music uses more than 100 scales, called ***maqam*** (ma-KAHM)—although only about 20 are normally played. Each *maqam* also has its own mood. Just as we think of minor scales as sad and major scales as happy and upbeat, Arab composers use a variety of scales to CONVEY a range of emotions.

Creative Changes

Arab musicians also express emotion through improvisation, playing around with a melody to create interesting, new musical lines. This makes their music both beautiful and unique because the music is performed differently every time it is played.

One big difference between Arabic and Western music is that Arabic music does not use harmony. This means chords are rarely played! Instead, Arab musicians make their melodies very long and complex—flowing and weaving up and down the scale. They also employ something called **heterophony,** in which a group of musicians will play the same melody slightly differently, either in a different rhythm or tempo, or with different EMBELLISHMENTS.

Explore!

With their special instruments and techniques, the musicians of the Arab world crank out as much rhythm as in our hip-hop, as much soul as in our jazz, and as much complexity as Western classical music. So open your ears and minds and explore the sounds of Arab music! ◎

listen to: "KEYF ANSAK (I CAN'T FORGET YOU)" BY ABDU EL-HANID

LESSON PLAN

Sounds of the Arab World

Objectives

- Learn about ethnomusicology and its instrument classification system through the study of Arab instruments (NS 6)
- Analyze the rhythm and melody of a classical Arab piece (NS 6)
- Perform rhythm patterns from Arab music (NS 2)

Materials

- *Music from Around the World* CD
- Student Worksheet: *Be an Ethnomusicologist* (page 44)
- Classroom percussion instruments

START

1 Have the class echo you as you chant the following syllables:

Once students feel comfortable with the vocal pattern, have them pat "dumm" on their right thigh and "takk" on their left thigh. Share that this is a rhythm pattern called *quddaam* (ka-DAHM) found in Arab music. Each pattern is described by the order of two different sounds:

low—*dumm* (represented in the notation by notes with downward stems)
high—*takk* (notes with upward stems)

When performing these on a drum, for example, musicians use the middle of the drum for the *dumm* and the edge of the drum to produce the *takk*. Let students try to transfer the rhythm to a hand drum or frame drum, playing the *dumm* sound as an open tone and the *takk* as a closed tone.

2 **LITLINK** Invite students to take turns reading "Sounds of the Arab World" on pp. 40–41 aloud. Check their understanding of the highlighted vocabulary words in the article. **[This reinforces vocabulary.]**

ORNATE—decorated in great detail
DIALECTS—varieties of a language
TERRITORY—geographical area
MESMERIZING—hypnotic, spellbinding
CONVEY—communicate
EMBELLISHMENTS—enhancements, adornments

DEVELOP

1 Introduce the field of study called ethnomusicology. Tell students that ethnomusicology is the study of music and people.

2 Share with students that when ethnomusicologists began collecting instruments from different countries, they needed to put them in categories to organize them. They found that the Western system of classification (strings, woodwinds, percussion, etc.) posed certain problems. For example, some instruments fell into two different categories.

ASK **Which instrument group does the piano belong to?** (Percussion)
Which other group could it belong to? (strings)

Because they needed to create categories that would work with all of the instruments of the world, ethnomusicologists set up classifications based on how instruments make sound.

3 Distribute copies of the Be an Ethnomusicologist worksheet. Ask students to complete the worksheet by classifying the instruments described in the article according to the categories given. Extend this activity to instruments you may have in the classroom or in the school orchestra or band.

4 Ask for a volunteer to state the difference between unison and hetereophony. (Hetereophony is when more than one instrument plays the same melody but each adds its own timbre, ornaments, and interpretations.)

5 Play the first two minutes or so of "Keyf Ansak" (CD track 12) by Abdul el-Hanid. Tell students that this piece is an example of ***raks sharqi*** (rocks SHARK-ee), or belly-dancing music. Ask students to listen to the different instruments they hear, write down the characteristics of each (string, drum, etc.), and attempt to identify them based on their descriptions in the article *(oud, dumbek, riqq, violin, qanun, ney).*

"Keyf Ansak" is a good example of heterophony. Play the following two sections of "Keyf Ansak" (Section I: 1:00–1:30; Section II: 3:25–3:55) and have students listen for the melody as sung by the male singer and played by the *qanun*.

ASK **What is different about the way the *qanun* plays in the second section?** (It is more ornamented and does not follow the vocal melody as closely.) **What does this tell you about heterophony?** (Players can change the melody a little bit every time.)

6 Introduce some other common Arab rhythms to your students. As in Western music, Arab music uses different rhythms, called *iqa'at* (ee-KAH-aht) [singular = *iqa*] for different pieces. Each of these rhythm patterns has a different name. Lead your class in trying the following three *iqa'at*. Have students chant "dumm" and "takk" in time first before attempting to play these rhythms on classroom instruments.

Wahdah wa Nusf [*WAH-dah wah NOO-suf*]

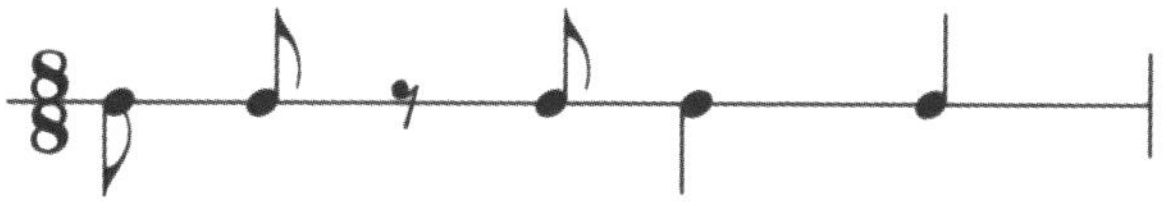

Sama'i [*sah-MAH-ee*]

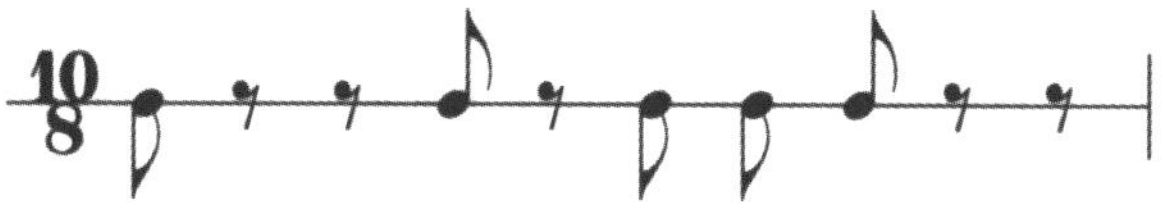

Wahdah Taqasim [*WAH-dah tah-KAH-seem*]
(usually performed at a walking tempo)

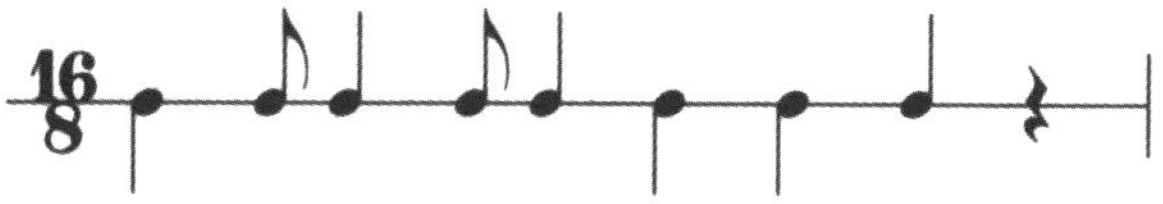

ASSESS **Particularly with *sama'i,* can students maintain the rhythm through several cycles?**

CLOSE

1 Play the first minute of "Keyf Ansak." The rhythm starts after a 25-second instrumental introduction.

ASK **To which category does the instrument playing belong?** (Chordophone)
Which Arab instrument do you think it might be? *(Oud)*
Which rhythm pattern of the four we learned is played in this song? *(Quddaam)*

2 Invite students to play the *quddaam* rhythm pattern along with the song.

Worksheet Answers
1. Aerophones—*ney,* clarinet; chordophones—*oud, qanun,* violin; idiophones—*riqq;* membranophones—*riqq, dumbek;* 2. no; 3. answers will vary; 4. answers will vary

Student Worksheet: Middle East

Name________________ Grade____ Music Teacher______________

Be an Ethnomusicologist

Ethnomusicologists use the following categories to classify the different instruments found all around the world:

- **Aerophones**—instruments in which sound is produced by a vibrating column of air (trumpet, flute, harmonica)
- **Chordophones**—sound is produced by the vibration of strings (guitar, violin, zither, harp)
- **Idiophones**—sound is produced from the vibration of the body of the instrument itself (wood block, clave, cymbals, bells)
- **Membranophones**—sound is produced by the vibration of a stretched membrane (mainly drums)

1. Fill in the chart below by deciding which of the instruments mentioned in the article belong in each category. Some may fall into more than one category.

Aerophones	Chordophones	Idiophones	Membranophones

2. Were there any categories that were not used? Yes No
If so, which one(s)? ______________________________

3. If you have classroom instruments, write these in their appropriate categories in the chart above.

4. Think of any other instruments not mentioned above and see if you can classify these correctly. For example, how would you classify the instruments played in your school orchestra?

Recommended Listening

Music of the Arab People

Rai

Cheb Mami ◎ ***Dellali*** (Ark 21)
Amr Diab ◎ ***Very Best of Amr Diab*** (EMI)
Rachid Taha ◎ ***Made in Medina*** (Ark 21)
Cheb Khaled ◎ ***Kutche*** (Stern's)
Samira Said ◎ ***Halal*** (Rotana)

Raks Sharqi

George Abdo ◎ ***Art of Belly Dancing*** (Monitor)
Rafat Misso & Hossam Ramzy ◎ ***Ahlamy*** (ARC)
Mokhtar al Said ◎ ***Raks Sharki: Belly Dance*** (Piranha)

Folk

Hamza El Din ◎ ***Music of Nubia*** (Vanguard)
Various Artists ◎ ***Sounds of Sudan*** (World Circuit)
Anouar Brahem ◎ ***Barzakh*** (ECM)

Classical

Simon Shaheen ◎ ***The Music of Mohamed Abd el-Wahaab*** (Axiom)
Umm Kulthum ◎ ***El Atlaal*** (Sono Cairo)
Fairuz ◎ ***The Very Best of Fairuz*** (Vox de l'Orient)

Sounds of India

In 1965, the Beatles released a song called "Norwegian Wood" that featured an Indian instrument called the **sitar** (SIH-tar). The shimmering melodies and echoes in this very popular recording gave many people their first taste of Indian music. But there's much more to the music of India than a single instrument—just ask any of the one billion people who live there. So let's tune in to India: a range of sounds and textures more VAST than the Himalayas is waiting to be explored. We'll just scratch the surface...

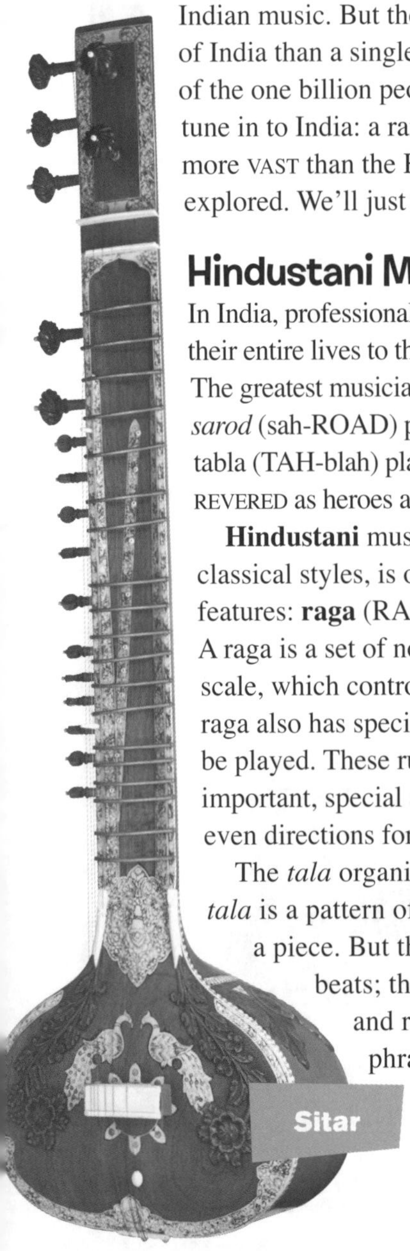

Sitar

Hindustani Music: A Classic

In India, professional classical musicians dedicate their entire lives to their music and their instrument. The greatest musicians—like sitarist Ravi Shankar, *sarod* (sah-ROAD) player Ali Akbar Khan, and tabla (TAH-blah) player Zakir Hussain—are REVERED as heroes and remembered as legends.

Hindustani music, one of the main Indian classical styles, is organized into two main features: **raga** (RAH-gah) and ***tala*** (TAH-lah). A raga is a set of notes, like a major or minor scale, which controls the melody. But each raga also has specific rules on how it should be played. These rules include which notes are important, special SEQUENCES of pitches, and even directions for when to sharp or flat a note.

The *tala* organizes the rhythm. Like meter, a *tala* is a pattern of beats that repeat throughout a piece. But these are not just identical beats; the patterns include accents and rests that suggest form and phrasing. A percussionist, like a tabla player, will blend different sounds into the pattern to keep things interesting.

In a performance, there is usually a melodic instrument, like a sitar or *sarod,* and a percussion instrument, like a tabla. The raga and *tala* INTERTWINE to build different emotional qualities into their performance, such as calm, tension, surprise, and release. The best musicians start with the basic raga and *tala,* then improvise to make something unique and special.

What to Play?

Consider a few of the most important Hindustani instruments:

The **sitar,** probably the most well-known of all Indian instruments, is related to the lute. It is wooden and features a very long neck strung with 18, 19, or 20 strings. A player holds the instrument in his or her lap and plucks or strums the strings with one hand. A shimmery echo sound comes from additional strings under the fretboard, which vibrate and ring "sympathetically" when certain pitches are played.

The **tabla** is a set of two round drums of different sizes and timbres. The drums are played with the fingertips and palms. Each hand plays one drum. Tabla drummers can make many different sounds by hitting, slapping, or rubbing the middle or edge of the drums. Each of these sounds has a special name, so drum patterns can be recited orally.

The **harmonium** is like an accordion that sits on the ground. Rather than squeezing it, however, a player pulls the back of the instrument in and out to make air flow and sound the pipes while playing a small attached keyboard with their other hand.

The ***sarod*** is smaller than a sitar but also a string instrument. Its wooden body is covered with a stretched skin. Like a violin, it has no frets on its chrome-plated neck so a player can easily slide notes up and down.

Hooray for Bollywood!

Which city makes the most movies in the world? If you said Hollywood, you're wrong! It is actually the Indian city of Mumbai. **Bollywood**—a play on Hollywood plus Mumbai's former name, Bombay—is

Tabla

the name for the Indian film industry.

Most Indian movies would seem like musicals to you. Actors sing and dance, twirling around the scenes in big, CHOREOGRAPHED numbers. Up until a few years ago, most Bollywood films were based on Indian myths or stories from history. Today, many feature Western stories. For example, *Bride and Prejudice* from 2004 was based on British writer Jane Austen's novel *Pride and Prejudice*.

The music is one of the main reasons why people go to Bollywood films—even though most of the songs are actually lip-synched! Professional singers record the songs in a studio, and these singers often become bigger stars than the lead actors. Asha Bhosle, one of the biggest film singers of all time, sang songs for young actresses well into her old age!

One Billion People Can't Be Wrong

Whether it's the classical Hindustani melodies of old, the popular Bollywood film music, folk music from the countryside, or the pop music of the cities, Indian music has spread across the globe. And it's no surprise: with its ancient traditions and modern styles, it offers beats and melodies that are truly unique.

A Sitar Legend's Legacy

Indian sitarist and composer Ravi Shankar is best known for his virtuosity on the sitar and his association with the Beatles. He was actually a mentor to George Harrison, who began learning the sitar in 1965. It was Shankar and the Beatles who were largely responsible for bringing Indian classical music to the attention of the West. With a career spanning six decades, Shankar holds the Guinness Record for the longest international career. He has written numerous compositions for sitar and other instruments, as well as music for films and ballets, and was a major influence on avant-garde composer Philip Glass, with whom he has collaborated. His accomplishments have been recognized with three Grammy Awards and an Academy Award nomination. Shankar's musical legacy continues with his children. His daughter Anoushka is also a world-famous sitarist and composer, and his other daughter, Norah Jones, has achieved massive professional success as a pop/jazz singer.

listen to: "RAGA MISHRA GHARA DHUN" BY RAVI SHANKAR AND "PASHTO" BY ZAKIR HUSSAIN

Sounds of India

Objectives

- Learn and perform three *tala*, or rhythm patterns, from Indian classical music (NS 2, 9)
- Demonstrate the principle of sympathetic string vibration (NS 8)
- Listen for and identify the *tala* in a recording of Indian classical music (NS 6)

Materials

- *Music from Around the World* CD
- Student Worksheet: *Musical Memory* (page 50)
- Two guitars, violins, or other stringed instruments

START

1. Begin class by teaching students one of the most basic *tala*, or rhythm patterns, from Indian classical music. It is called *tintal* (teen-tahl). Without introduction, have students repeat the following syllables after you:

   dhaa (dah) *dhin* (deen) *tin* (teen) *naa* (nah)

2. Now teach students the *tintal* pattern below, which contains a total of 16 syllables. (Read this pattern from left to right, top to bottom.) Start by repeating the first row only. Maintain a steady beat at a comfortable walking tempo (76 bpm). Then add the second row to the first and repeat only the first and second rows. Continue the same way with the third and fourth rows until students can repeat the entire pattern.

^{x}dhaa	dhin	dhin	dhaa
^{2}dhaa	dhin	dhin	dhin
^{0}dhaa	tin	tin	naa
^{3}naa	dhin	dhin	dhaa

3. Once students have memorized the vocal pattern, have them add hand claps and a hand wave at the beginning of each four-beat section. Clap when saying a syllable that has an x, 2, or 3 in front of it. Wave one hand (start palm down and turn it over) when saying the syllable that has a 0 in front of it. Continue until students can repeat the entire pattern without error.

4. Explain that this *tintal* rhythm pattern is divided into sections *(vibhaj)*, which are just like the measures that are used to divide European and American music.

   ASK **What do you think divides the pattern we just learned and shows where the sections are?** (The claps and the waves)
   How many beats are in each section? (4)

DEVELOP

1. Now write the following *jhaptal tala* on the board as it is shown below. (Read this pattern from left to right, top to bottom.)

^{x}dhin	naa	
^{2}dhin	dhin	naa
^{0}tin	naa	
^{3}dhin	dhin	naa

Share with students that a *tala* does not need to be symmetrical, or divided into sections of equal length, like the *tintal* pattern they just learned. Point to each syllable as you say it on a steady beat.

ASK **From looking at this *tala*, what do you notice is different or similar?** (It isn't as long; it uses the same syllables; it is broken into uneven sections.)

Now say and clap/wave this pattern as you did with the *tintal* pattern above and have students repeat after you.

2 **LITLINK** Have the class read "Sounds of India" on pp. 46–47 either silently or softly aloud in pairs, depending on their abilities. Check students' understanding of the highlighted vocabulary words in the article. **[This reinforces vocabulary.]**

VAST—extremely large; huge
REVERED—held in great respect
SEQUENCES—repeating similar patterns of notes
INTERTWINE—mix together
CHOREOGRAPHED—composed dance steps

3 **LITLINK** Distribute copies of the Musical Memory worksheet and give students time to complete it. Review the answers together as a class. **[This reinforces comprehension.]**

4 Tell students they are now going to hear a piece by Ravi Shankar, one of the greatest sitar players of all time. Play "Raga Mishra Ghara Dhun" (CD track 13) and have students listen for a minute.

ASK **How many instruments did you hear?** (Two; sitar and tabla) **What types of instruments did you hear?** (Stringed, drum) **What roles do you think they were playing?** (The stringed instrument was soloing, the drum was accompanying.)

Explain that the sitar is improvising over the repetitive *tala* rhythm of the tabla.

5 Now ask students to listen for the buzzing string sound. Share that these are the **sympathetic** strings of the sitar. Sympathetic strings are extra strings that run under the fretboard; they are never strummed, but vibrate whenever the same pitch is played on the main strings on top.

6 Explain that this acoustic phenomenon can be reproduced with any two pitched instruments. You can demonstrate this for students with two guitars or violins that are tuned precisely to each other. Hold the instruments close to each other and pluck a string on one. Then stop this string and see if the same string on the other instrument is vibrating. This is called sympathetic vibration. Allow students to experiment and demonstrate.

CLOSE

1 Now play the first 45 seconds of "Pashto" (CD track 14), performed by tabla player Zakir Hussain. Share that Hussain is one of the most respected tabla players in the world today.

Explain to students that *pashto* is the name of another rhythm pattern. Write the pattern shown below on the board. Tell students that this *pashto tala* is 7 beats long, and that these beats are divided into sections of 3 + 2 + 2.

^{0}tin	—	—
^{1}dhin	—	
^{2}naa	naa	

2 Teach students the *tala* above. (Note: This version of *pashto* is slightly modified to make it easier for students to perform. Rest on the boxes with dashes.)

3 Now play "Pashto" (CD track 14) again. Starting at 1:40, have students count in their heads and try to identify the number of beats in this asymmetrical *tala*. (Tell them to listen to the ostinato of the *sarod*, the instrument playing the melody, which indicates the beats.)

Worksheet Answers
1. i); 2. c); 3. f); 4. j); 5. l); 6. k); 7. h); 8. b); 9. d); 10. e); 11. a); 12. g); the Beatles; Bollywood or Mumbai; sympathetic vibration

Student Worksheet: India

Name______________ Grade____ Music Teacher__________

Musical Memory

Can you remember all the instruments, artists, and musical concepts you read about in "Sounds of India"? Match the words and phrases in the first column to the appropriate description in the second column.

1. harmonium	a) a fretted instrument played by plucking or strumming strings
2. raga	b) a set of two small, round drums of different sizes and timbres
3. *sarod*	c) a set of notes (like a scale) used in Indian music
4. Bollywood	d) a set of rhythm patterns used in Indian music
5. Ali Akbar Khan	e) a famous sitarist and composer who holds the Guinness record for longest international career
6. Norah Jones	f) an unfretted string instrument that allows sliding pitches
7. Hindustani	g) a famous and revered tabla player
8. tabla	h) one of the main Indian classical styles
9. *tala*	i) an accordion that sits on the ground
10. Ravi Shankar	j) the name given to the film industry in Mumbai, India
11. sitar	k) daughter of Ravi Shankar who is a pop/jazz singer
12. Zakir Hussain	l) a famous *sarod* player

Bonus Questions

Which popular band visited Ravi Shankar in India and incorporated the sitar into their music?

____

Which city in the world produces the most movies? ____________________________

The phenomenon of one string vibrating at the same speed as another string is called __________

____

Recommended Listening

Indian Music

Classical

Zakir Hussain ◎ *Making Music* (ECM Records)

Ravi Shankar ◎ *The Essential Ravi Shankar* (RCA)

Ali Akbar Khan ◎ *Signature Series, Vol. 2: Three Ragas* (Ammp)

Purbayan Chatterjee ◎ *New Horizons* (Rhyme)

Rahul Sharma ◎ *Maya the Illusion* (Navras)

Zia Mohiuddin Dagar ◎ *Raga Yaman* (Nimbus)

Pramita Mallick ◎ *Chhandey Nanan Rang* (Rhyme)

Ritesh Das ◎ *Weaving* (NAXOS)

Nancy Lesh ◎ *Cello* (India Archive Music)

Soundtracks/Pop

Various Artists ◎ *Salaam Namaste* (Yah Raj)

Various Artists ◎ *Kal Ho Naa Ho* (Sony)

Various Artists ◎ *Lagaan: Once Upon a Time in India* (Sony)

Various Artists ◎ *Namastey London* (Eros Music)

Asha Bhosle ◎ *The Rough Guide to Asha Bohsle* (World Music Network)

Various Artists ◎ *The Rough Guide to the Music of India* (World Music Network)

Bally Sagoo ◎ *On the Mix: The Story So Far* (Star)

Talvin Singh ◎ *OK* (Island)

Sounds of China

China is one of the oldest civilizations in the world, and is the source of "the four great inventions": the compass, gunpowder, paper, and printing. The Chinese also invented many musical instruments, which make exotic sounds quite unlike what we hear on the radio. But pop and rock music, with their familiar sounds of electric guitars, drums, bass, and keyboards, have become a huge part of Chinese culture as well. The music of Chinese artists such as rock legend Cui Jian (tsway JEE-en), sweet-voiced pop star Miriam Yeung, and hard rockers Hei Bao perk up even the sleepiest Chinese teenager. This huge country creates a large range of music, making the sounds of China a mix of old and new.

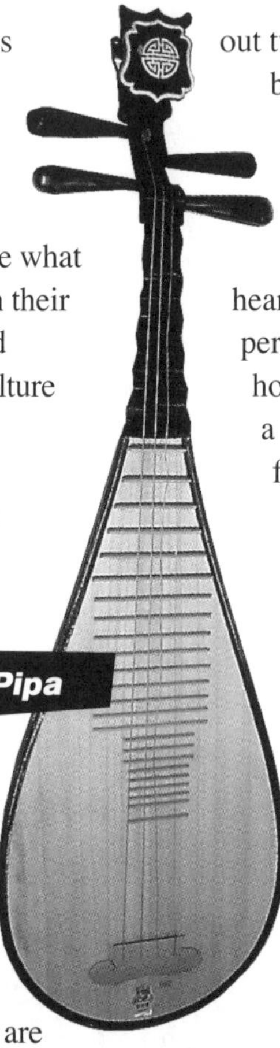

Pipa

Time-Honored Traditions

The Chinese culture reaches back to 2000 B.C.—more than 4,000 years ago. It was one of the earliest civilizations in the world, so it's not surprising that the Chinese take a lot of pride in their traditions and their music.

The Han Chinese make up over 90 percent of the population. While their traditional music styles are extremely DIVERSE, they do share some common elements, such as heterophony. **Heterophony** is when different instruments play the same melody but each musician adds his or her own personal touches, like extra notes or rhythmic changes. Another common feature of many Chinese music styles is the pentatonic scale (**pentatonic** means "five tones"). If you play only the black keys the next time you're at the piano, you'll hear the pentatonic scale.

Soft and Loud

In southeastern China, "silk and bamboo" music called ***sizhu*** (SUH-droo) can be heard in teahouses, China's version of American coffee shops. Like in a mini–jam session, everyone brings an instrument, and people call out tunes for the musicians to play. The name "silk and bamboo" comes from the materials some of the instruments are made of, such as the silk strings of the ***erhu*** or the bamboo ***dizi*** (see the sidebar, "Chinese Instruments").

In many northern Chinese villages, you'll hear the incredibly loud sound of drums and other percussion instruments accompanying oboe-like horns. These horns, called ***suona*** (so-nah), sound like a long, loud duck quack! *w* players are also known for their crazy THEATRICS.

In the cities, many people go to the opera—but it isn't to see Placido Domingo or *Carmen*. China has its own operas that are full of its history and stories. The scenery, acting, and costumes are as fancy as anything you'd see at the Metropolitan Opera in New York, but the singing can sound a little odd to non-Chinese. Often, the singers use tense, GUTTURAL sounds, or strain to sing very high, almost whining notes.

Get Off My Back!

The Chinese government began to seriously MEDDLE in China's music in the 1950s. After the civil war in the 1940s, the communists took over the country, and for much of their rule they pressured musicians to only compose and perform songs that supported the government. Imagine President Bush demanding that the Black-Eyed Peas or Gwen Stefani sing a song called "Long Live George"!

From 1966 to 1976, music performances were strictly controlled by the government. It wasn't until after this period that Chinese musicians felt free to create their own music. The first major music star to EMERGE from these dark times was Teresa Teng, who was brave enough to record her albums independently and distribute them behind the scenes to avoid the authorities. Her spirit and songs ignited a pop-music explosion that inspired other musicians to break out in the early 1980s.

Rock Revolt

Though the Chinese still enjoy traditional music, most young people prefer to listen to pop music. **Cantopop,** which started in Hong Kong, has a sound you would probably recognize. Most of its songs are about love and relationships, and singers are backed by guitars, synthesizers, and drum kits. Many Cantopop musicians, like Joey Yung, Twins, and Miriam Yeung, are huge stars with gold records, their own websites, and millions of screaming fans.

For many years, Cantopop was pretty much the only pop music in China. Then along came Cui Jian, a VISIONARY who played trumpet and guitar and tore the roof off the local music scene. A rock music pioneer like the Rolling Stones or Led Zeppelin in the U.S., Cui combines elements from different music styles, including rock, rap, reggae, and even traditional Chinese. And like a true revolutionary, he doesn't always sing what the government likes—he stands up for what he believes in.

Looking Forward

Inspired by Cui, more and more young Chinese musicians are starting to play rock music and gaining popularity. Bands like Tongue, Happy Avenue, and Wood Pushing Melon are rocking in the capital city of Beijing. Before long, all of China will be joining the musical revolution! ◎

CHINESE INSTRUMENTS

ERHU (ur-hoo) This two-string fiddle rests on the player's knee instead of under the chin.

PIPA (pee-pah) This 4- to 5-string lute is held upright like a small cello, so the neck rests on the player's shoulder.

DIZI (dee-zee) This bamboo ute has a very thin reed or bamboo membrane that buzzes when played, making a shrill sound like a kazoo.

YANGQIN (yahn-chin) This is a type of hammered dulcimer that looks like the inside of a piano. Players hit the strings with small, thin mallets to make the strings ring.

SHENG (shung) A group of small bamboo pipes bound in a tight circle and connected to a base. Players blow into a hole in the base to sound the pipes.

listen to: TWO TRADITIONAL PIECES: "THUNDER IN THE DROUGHT" AND "MOON IN THE SKY"

LESSON PLAN

Sounds of China

Objectives

- Discuss instrument classification systems (NS 8, 9)
- Identify instruments in Chinese music (NS 6)
- Identify performance/ articulation techniques for string instruments (NS 6)
- Identify heterophony in a Chinese folk song (NS 6)
- Notate a Chinese scale (NS 5)
- Create a pentatonic ostinato pattern (NS 1, 4)

Materials

- *Music from Around the World* CD
- Classroom instruments
- Student Worksheet: *Listen and Learn* (page 56)
- Staff notation paper

START

1. Before students arrive, place about 10 classroom instruments (or as many as you have) at the front of the room (e.g., wood block, shaker, drums, guitar, xylophone, flute, etc.). Call on volunteers to identify the instrument category names for the instruments (e.g., strings, woodwinds, etc.). Ask students to consider each instrument's shape, method of sound production, material, and pitch. Based on this, discuss which category names make sense.

2. Explain that instrument categories can be called "classification systems." Share that in ancient China, instruments were grouped into eight categories based on the material that was used to make them. Using the chart below, write down each of the eight Chinese categories with their corresponding materials on separate pieces of paper and tape to the wall or blackboard at the front of the room. Invite volunteers to select an instrument and place it under the category in which they think it belongs. Have them explain their reasoning.

Chinese Instrument Category	Type of Material
jin	metal
shi	stone
tu	earth (or clay)
ge	animal skin (membrane)
si	silk
mu	wood
pao	gourd
zhu	bamboo

3. Divide the class into pairs, and ask students to write the eight categories on a sheet of paper. Challenge them to think of at least two additional instruments that might fit in each category. Invite volunteers to share and explain their choices to the class.

DEVELOP

1. Have students read "Sounds of China" on pp. 52–53 either silently, softly aloud in pairs, or as a class, depending on their abilities.

   LITLINK Review the highlighted vocabulary words in the article. Call on students to try to determine the meaning of each word from its context. **[This reinforces vocabulary.]**

DIVERSE—varied, different
THEATRICS—playful actions
GUTTURAL—(sound) made from the back of the throat
MEDDLE—interfere with; become involved in
EMERGE—come onto the scene
VISIONARY—person who is forward-thinking

2. Distribute copies of the "Listen & Learn" classroom worksheet. Share with students that the song they are going to hear, composed in the early 1900s, is an example of Cantonese teahouse music. Tell them that this music style is very similar to the "silk and bamboo" music described in the article.

3. Play "Thunder in the Drought" (CD track 15). Ask students to try to identify some of the instruments in the song based on their descriptions in the article. (Instruments include *erhu*, *pipa*, *dizi*, and *yangqin*, as well as a drum, cymbal, and wood block.)

15

Allow students the opportunity to listen to "Thunder in the Drought" as many times as they need in order to complete the worksheet. You may want to stop and start the recording so that you can answer the questions together.

4. Now play "Moon in the Sky" (CD track 16), a piece for solo *pipa* performed by world-renowned player Wu Man. Explain to students that *pipa* players wear hard, fake fingernails on the thumb and fingers of their right hand, which they use to pluck the strings, sometimes very rapidly. Players use their left hand to create many of the interesting effects heard in this track.

16

5. Ask students to listen to and try to describe the *pipa* player's different techniques. If you have a violin or guitar in the classroom, you can demonstrate some of the following techniques to the class:

- **Bending**—pulling the string with a left-hand finger to change the pitch
- **Tremolo**—repeatedly plucking the same string with the right hand fingers
- **Slides (or glissando)**—sliding a left finger along the string after it is plucked
- **Vibrato**—pulling a string slightly back and forth with the left hand after it is plucked
- **Harmonics**—lightly touching a string at specific intervals (such as half its length) and plucking to produce high-pitched overtones (used around 1:36)

Explain that rock, jazz, and blues guitarists often use these techniques in their solos. (You can play an example of a blues or rock guitar solo so students can hear the similarities.)

6. Play "Thunder in the Drought" (CD track 15) again and have students focus on what the different instruments are playing.

15

ASK **Do the various instruments play exactly the same thing?** (Not exactly, but very similar)
What is this feature called? (Heterophony)

7. Distribute staff paper to the class. Tell students that this piece uses a G pentatonic, or five-note scale, which includes the notes G-A-B-D-E. Guide students in notating these pitches on staff paper (on a treble clef).

Review the intervals between each pair of notes in the scale. Then ask students to notate pentatonic scales in the keys of C, D, and F using those same intervals. (As an alternative or extension, ask them to create these pentatonic scales using a mallet instrument such as an alto xylophone.)

CLOSE

1. Set up mallet instruments in G pentatonic and let students practice playing the scale. (You may also use recorders, keyboards, or other instruments.)

2. Divide students into small groups. Have one or two members in each group create and play a 4- or 8-beat ostinato pattern on drum, wood blocks, or other classroom percussion instruments. Have the remaining members play a melody in the G pentatonic scale with the ostinato. Encourage students to create a fast-slow-fast form similar to what they heard in "Thunder in the Drought."

Worksheet Answers
1. 3); 2. a); 3. a); 4. b); 5. a bow; 6. b); 7. c); 8. c); 9. b); 10. true

Student Worksheet: China

Name________ Grade __ Music Teacher ____________

Listen & Learn

Listen carefully to "Thunder in the Drought," played by a small orchestra of traditional Chinese instruments. Review the different types of instruments outlined in the article and then fill in the blanks or choose the correct answer for each question below. You may need to listen to the piece more than once to check your answers!

1. At the very beginning of the piece, there are _____ drum beats.

2. The full ensemble usually plays:
a) in unison
b) two different melodies
c) in four-part harmony

3. When the *dizi* plays solo (1:35–1:42), I also hear:
a) triangle
b) low drum beats
c) wood block

4. The tempo of the piece
a) never changes
b) gets slower in one place and faster in another
c) gradually gets faster

5. An *erhu* is played with _________.

6. Most traditional Chinese music is based on:
a) a major scale
b) a pentatonic scale
c) no scale—it is atonal

7. Stringed instruments that are plucked and percussion instruments that are struck cannot sustain a tone like a wind instrument can. In this piece, how do they sustain a note for a long time?
a) By having a wind instrument play the note for them
b) By using a special amplifier
c) By playing the same note many times

8. When I listened for a bass line, I heard:
a) a lot of low-pitched instruments
b) sustained, long notes
c) almost no low notes

9. Most of the instruments in this ensemble play:
a) mostly in a low register
b) mostly in a high register

10. At the end of the piece, the ensemble plays the same melody it played at the beginning.

True False

Chinese Music

Traditional/Classical

Wu Man ◎ ***Chinese Music for the Pipa*** (Nimbus)

Various Artists ◎ ***Classical Chinese Folk Music*** (ARC)

Chinese Instrumental Ensemble ◎ ***Masterpieces of Chinese Traditional Music*** (Solar Music)

Various Artists ◎ ***China: Classical Music*** (Ocura)

Various Artists ◎ ***China: Time to Listen*** (Elipsis Arts)

Li Xiangting Chine ◎ ***L'Art du Qin*** (Ocora)

Various Artists ◎ ***The Hugo Masters: An Anthology of Chinese Classical Music, Vols 1-4*** (Celestial Harmonies)

Cantopop

Faye Wong ◎ ***Double Best: 1989-1997*** (PolyGram)

Hacken Lee ◎ ***Seoul Concert Hall, Vol. 2*** (PolyGram)

Joey Yung ◎ ***Give Love a Break*** (Emperor Entertainment Group)

Miriam Yeung ◎ ***Unlimited*** (Gold Label)

Opera

Various Artists ◎ ***Chinese Opera: Songs and Music*** (Smithsonian Folkways)

Various Artists ◎ ***Sichuan Opera: The Legend of White Snake*** (Buda)

Rock

Cui Jian ◎ ***Best of Cui Jian: 1986-1996*** (EMI)

Various Artists ◎ ***Rough Guide to the Music of China*** (World Music Network)

Glossary

The following terms appear in bold when introduced in the pages of *Music from Around the World*.

Aerophones: Wind instruments in which air is the primary means of vibration and sound. Examples include flutes and woodwinds, accordions and bagpipes, etc.

Apartheid: An oppressive system of government that ruled in South Africa from 1948 until 1994.

Arpa: A 32- to 36-string harp that plays bass and melody in Mexican *son jarocho* music.

Augmented second: An interval of three half-steps.

Bachata: A Dominican dance- and guitar-based song style often featuring tales of heartbreak and sadness. Made popular worldwide by Juan Luis Guerra.

Bairros: The neighborhoods in Brazilian cities.

Banda: A brass-based form of traditional music from northern Mexico.

Berimbau: A bow-shaped instrument featuring a steel string and a gourd resonator.

Bollywood: An informal name for the Hindi-language film industry based in Mumbai, India.

Bossa nova: Swinging style of Brazilian music created in the late 1950s; means "the new beat."

Caixa: Snare drums played in samba ensembles during Brazilian Carnaval.

Calypso: An Afro-Trinidadian song form often featuring lyrics about daily life.

Cantopop: A sub-genre of Chinese pop music; short for "Cantonese popular music."

Capoeira: Afro-Brazilian martial art/dance form developed in the 1500s by African slaves. Typically accompanied by *berimbau,* percussion, and chant.

Cavaquinho: A Portuguese instrument resembling a ukelele.

Chordophones: Instruments in which strings are stretched between fixed points and struck, plucked, or bowed to produce sound. Examples include harps, guitars, pianos, violins, etc.

Comparagga: A popular Haitian style of music that blends Jamaican reggae and *compas.*

Compas: Haitian dance music developed by bandleader Nemours Jean-Baptiste in the 1950s; originally called *compas-direct.*

Conga: A barrel-shaped, single-headed hand drum from West Africa. A mainstay of Latin Caribbean percussion.

Corrido: A Mexican ballad that describes life in a border town.

Cumbia: A Latin style of song and dance played in 4/4 time with a heavy beat on 1 and accentuated beats on 3 and 4; often classified as salsa.

Cymbalom: A stringed instrument that is played with mallets in *klezmer* music. Also known as a hammered dulcimer. See also *tsimbl.*

Dancehall: A type of Jamaican reggae characterized by a DJ singing and rapping (or "toasting") over raw reggae tracks.

Diaspora: The migration of the Jewish people across parts of Europe, Asia, and North Africa.

Dizi: A Chinese bamboo flute containing a thin reed or bamboo membrane that buzzes when the instrument is played.

Dub: A stripped-down reggae mix incorporating echo and other sound effects.

Dumbek: A goblet-shaped drum from the Middle East and North Africa made from clay, wood, or metal, with a goatskin head; traditionally held under the arm when played.

Erhu: A Chinese two-string fiddle.

Favelas: The slums in Brazilian cities.

Forró: A hard-driving music style from northeast Brazil featuring accordion.

Güira: A serrated gourd scraped with a stick. Played throughout the Caribbean.

Guitarrón: A large, deep-bodied, fretless 6-string acoustic bass played in Mexican mariachi bands.

Harmonium: An instrument used in Indian music that's similar to an accordian, but freestanding. A player pulls the back in and out to make air flow and sound the pipes while playing a small attached keyboard with the other hand.

Heterophony: Simultaneous executions of the same melody that differ in either instrument played, texture, tone, or subtle elements of performance.

Hindustani classical: The classical music of northern India.

Idiophones: Instruments which produce sound from the vibrations of their substance, without the use of strings or membranes. Examples include bells, rattles, triangle, etc.

Isicathimiya: An all-male *a cappella* choral singing style from South Africa that features call-and-response. Popularized internationally by Ladysmith Black Mambazo.

Jarana: A small, guitar-shaped, fretted instrument with 8 nylon strings from the Veracruz region of Mexico. Used in *son jarocho* music.

Jive: South African urban electric pop music; also known as township jive. See *kwela*.

Klezmer: Jewish secular music that emerged in Eastern Europe during the 15th century; characterized by clarinets, fiddles, and chromatic melodies.

Kwaito: A South African urban genre developed during the 1990s that blends South African dance music, hip-hop, Jamaican influences, house music, and R&B.

Kwela: Penny whistle street music that emerged in South Africa during the 1950s; developed into township jive.

Maqam: Any of approximately 120 scales using the characteristic quarter tones of Arab classical music and its popular descendants.

Marabi: An early music/dance form that arose in the slums of Johannesburg during the 1920s and '30s; a blend of early African-American Dixieland and ragtime with cyclical harmonies and a trance-like rhythmic foundation.

Mariachi: A musical style originating in the Mexican state of Jalisco. Ensembles feature 3–20 musicians playing violins, trumpets, guitar, *vihuela*, and *guitarrón*. Performers dress in silver-studded *charro* outfits with wide-brimmed hats.

Marimba: An idiophone of Southeast Asian or African origin sounded by striking rosewood bars laid over brass resonating tubes.

Maghreb: The North African part of the Arab world (excluding Egypt and the Sudan).

Mashreq: The Asian part of the Arab world.

Mbaqanga: South African township music popular from the mid-1960s to the mid-1970s.

Mbira: A hand-held instrument from Africa featuring a gourd or wood resonator attached to tuned metal keys that are plucked by the thumbs and right index finger; also known as a thumb piano.

Membranophones: Instruments featuring a membrane stretched across a hollow cylinder that produces sound when it is struck or rubbed (e.g., drums); the membrane is usually made of animal hides or synthetic material.

Mento: Jamaican folk music that mixes the music traditions of the English who colonized the country with those of the African slaves brought there.

Merengue: Folk and popular music of the Dominican Republic played in a fast 2/4

Glossary continued

meter; essential instruments are *tambora* and *güira*, with *congas* added in modern bands. Originally featured accordion; today's bands include keyboards and saxophones.

Merenhouse: A Caribbean music style that blends *merengue* with contemporary house music sounds.

Merenrap: A Caribbean music style that blends *merengue* with the sounds of contemporary rap.

MPB: *Musica popular brasileira*, or Brazilian popular music.

Ney: Reed flute popular throughout the Middle East, with origins in ancient Egypt.

Norteño: A traditional style of music based largely on *corridos*, *rancheras*, and polka that originated in rural northern Mexico and Mexican communities in the U.S. border states in the early 20th century; literally "northern" in Spanish.

Oud: An Arab lute consisting of a large wooden sound box, a short fretless neck, and usually six paired strings with a single bass string.

Pentatonic (scale): A five-tone scale commonly featured in Chinese music.

Pipa: A pear-shaped Chinese lute with four strings.

Qanun: A Middle Eastern zither featured prominently in Arab classical music. A predecessor of the harp and piano.

Raga: A melodic framework for improvisation in Indian classical music based on a set of notes (usually five to seven) and characteristic rhythmic patterns.

Rai: A popular youth-oriented Arab music style that originated in Algeria characterized by political themes and a vocal style that carries single syllables over different notes.

Raks sharqi: Arab belly-dance music originating in Egypt.

Rapso: Trinidadian music that mixes *soca* with rap.

Rastafarianism: A back-to-Africa cultural movement and religion begun in Jamaica during the 1930s that reveres Emperor Haile Selassie of Ethiopia and provides the spiritual basis of reggae music.

Reggae: Jamaican music style featuring a strongly accented offbeat. Evolved from the older styles of *ska* and *rocksteady* and led to related styles such as *dub* and *ragamuffin*; themes include politics of the poor and the Rastafarian religion. Popularized by Bob Marley and Peter Tosh.

Requinto: A Mexican 6-string nylon guitar that is smaller than a standard guitar and has a deeper body.

Riddims: Repeating rhythms used in Jamaican dancehall.

Riqq: A percussion instrument of the Arab world similar to a tambourine and featuring a head made of fish skin.

Rock en Español: A modern Mexican music style influenced by American rock.

Rocksteady: A Caribbean musical style that shares elements of both reggae and *ska* but is performed at half the tempo of *ska*.

Samba: The most popular music style in Brazil, featuring a characteristic "one-two" beat. Widely viewed as the national music of Brazil.

Samba-canção: Melody-driven Brazilian samba where a singer is accompanied by a guitar and a *cavaquinho*.

Samba-enredo: Theme samba developed by an *escola de samba* and performed by large ensembles during Carnaval celebrations in Brazil.

Samba-reggae: Brazilian version of reggae developed in Salvador de Bahia, in northeastern Brazil, by the group Olodum.

Sarod: A 25-string lute carved from a single block of wood, with a steel fretless fingerboard. Common in Hindustani classical music.

Sesquialtera: A shifting meter featured in many Mexican music styles.

Sheng: A Chinese instrument consisting of a group of small bamboo pipes that are bound in a tight circle and connected to a base; a player blows into a hole in the base to sound the pipes.

Sitar: A long-necked Indian lute that usually features 17 strings: 6–8 are for playing and droning, the rest are sympathetic strings that resonate under the frets. Common in Hindustani classical music.

Sizhu: "Silk and bamboo" music heard in teahouses of southeastern China.

Ska: Uptempo, horn-based instrumental music that originated in Jamaica in the early 1960s. Revived as part of the "two-tone" movement in the U.K. in the early 1980s with the English Beat, and evolving with modern *ska rock* bands Smash Mouth and No Doubt.

Soca: A music style with a heavy bass-drum beat and prominent cymbals that originated in Trinidad and Tobago and is a variant of calypso; name derived from "the (so)ul of (ca)lypso."

Soca chutney: The *soca* music of the Indian population in Trinidad.

Son istmeno: A music style transported from West Africa to southern Mexico in which the *marimba* is prominently featured.

Son jarocho: Harp music from the Veracruz region of Mexico featuring a highly repetitive musical structure and improvisational lyrics.

Steel drum/pan: A pitched percussion instrument from Trinidad made from recycled 55-gallon oil drums; multiple steel drums are tuned chromatically and often played as a set, providing both percussive and melodic effects.

Suona: A high-pitched traditional Chinese double-reed instrument similar to an oboe.

Surdo: A very deep, loud bass drum essential to Brazilian *samba*.

Tabla: A pair of small drums: the higher-pitched *daya*, played with the right hand, is a cylindrical wooden one-skinned drum tuned to the raga's tonic; the *baya*, played with the left hand, is a deep copper kettledrum whose pitch varies according to the pressure from the heel of the player's hand. The principal percussion in Hindustani classical music.

Takhts: Small ensembles that play both the classical and folk music of the Arab world.

Tala: The Indian system of rhythm.

Tambora: A two-headed goatskin drum that provides the characteristic *merengue* beat; held in the player's lap, one head is played with a stick and the other with the hand.

Timba: Long drums played with the hand or a light stick; used in Brazilian samba.

Tropicália: A musical/cultural movement that originated in Brazil during the 1960s with Gilberto Gil, Caetano Veloso, and others; the music combines samba, bossa nova, rock, and folk sounds.

Tsimbl: A stringed instrument that is played with mallets in *klezmer* music. Also known as a hammered dulcimer. See also *cymbalom*.

Twoubadou: A Haitian style of music similar to American R&B.

Vihuela: A high-pitched, five-string guitar played in mariachi bands.

Yangqin: A Chinese hammered dulcimer played with mallets.

CD Recording Credits

Track 1 **Mahotella Queens, "Women of the World"**
(E. Butler, M. Pilot)
Recording © 1994 Courtesy of Gallo Record Company. Published by Mike Pilot Productions. Used by permission. All rights reserved.

Track 2 **Soweto Gospel Choir, "Thina Simnoqobile"**
(G. Vilakazi, N. Vilakazi)
© 2005 Courtesy of Shanachie Entertainment Corp. Published by Isicholo Entertainment. Used by permission. All rights reserved.

Track 3 **Siquisirí, "La Guanabana"** (Traditional)
© 2005 Courtesy of ARC Music International.

Track 4 **Mariachi Los Comperos, "Los Arrieros"**
(S. Vargas, R. Fuentes)
© 2002 Courtesy of Smithsonian Folkways Recordings. Published by Peer International Corp. Used by permission. All rights reserved.

Track 5 **Luiz Bonfá, "Uma Prece"**
(L. Bonfá)
© 2005 Courtesy of Smithsonian Folkways Recordings. Published by Peer International Corp. Used by permission. All rights reserved.

Track 6 **Wilson Moreira, "Peso No Balança"**
(W. Moreira)
© 1989 Courtesy of Rounder Records. Published by EMI Music Publishing. Used by permission. All rights reserved.

Track 7 **Judy Mowatt, "Let's Dance"**
(S. Davis)
© 1987 Courtesy of Shanachie Entertainment Corp. Published by Sangie Davis Music. Used by permission. All rights reserved.

Track 8 **Calypso Rose, "Ju Ju Warrior"**
(M. Lewis)
© 1993 Courtesy of Ice Records, Ltd. Published by Ice Music, Ltd. Used by permission. All rights reserved.

Track 9 **Lord Kitchener, "Miss Tourist"**
(A. Roberts)
© 1994 Courtesy of Ice Records, Ltd.
Published by Ice Music, Ltd. Used by permission. All rights reserved.

Track 10 **Klezmatics, "Kolomeyke"**
(F. London)
© 2002 Courtesy of Rounder Records. Published by Psycho Freylekhs Music. Used by permission. All rights reserved.

Track 11 **Klezmatics, "Dybbuk Shers"**
(A. Svigals)
© 2002 Courtesy of Rounder Records.Published by Alicia Svigals Music. Used by permission. All rights reserved.

Track 12 **Abdu El-Hanid, "Keyf Ansak (I Can't Forget You)"**
© Courtesy of Smithsonian Folkways Recordings.

Track 13 **Ravi Shankar, "Raga Mishra Ghara Dhun"**
(R. Shankar)
© 1999 Courtesy of Music Today (IPRS).
Published by Music Today (IPRS). Rights for Music Today sub-published in the United States by Cherry Lane Music Publishing Company, Inc. Used by permission. All rights reserved.

Track 14 **Zakir Hussain, "Pashto"**
(R. Shankar)
© 2005 Courtesy of Music Today (IPRS). Published by Music Today (IPRS). Rights for Music Today sub-published in the United States by Cherry Lane Music Publishing Company, Inc. Used by permission. All rights reserved.

Track 15 **Asian Music Ensemble of Northern Illinois University, "Thunder in the Drought" (Han Tian Lei)**
(Y. Laolie)
© 1981 Courtesy of Smithsonian Folkways Recordings.
Published by CRC Jianian. Used by permission. All rights reserved

Track 16 **Wu Man, "Moon in the Sky" (Xiao Yue Er Gao)**
(P. Xiuwen)
© 2002 Courtesy of Smithsonian Folkways Recordings.
Published by CRC Jianian. Used by permission. All rights reserved.

World Music Sources

Record Labels and Online Music Retailers

Ace Records
www.acerecords.co.uk

Alula Records
www.alula.com

ARC Music
www.arcmusic.co.uk

Arhoolie Records
www.arhoolie.com

Ice Records
www.icerecords.com

Luaka Bop
www.luakabop.com

Music of the World
www.musicoftheworld.com/asia.html

Putumayo World Music
www.putumayo.com

Rounder Records
www.rounder.com

Shanachie Entertainment
www.shanachie.com

Smithsonian Folkways
www.smithsonianglobalsound.org

Stern's Music
www.sternsmusic.com

World Music Store
www.worldmusicstore.com

World Instrument Retailers

Apollo's Axes
www.apollosaxes.com

Berkeley Musical Instrument Exchange
www.berkeleymusic.com

Khazana
www.khazana.com

Lark in the Morning
http://larkinthemorning.com

Mideast Ethnic Music Instruments
www.mid-east.com/index.asp

Musantiqua Inc.
www.musantiqua.ca/mus/en/index.html

Rhythm Fusion
www.rhythmfusion.com

Books

Garland Encyclopedia of World Music, Vols. 1-10 (Routledge, 1999)

Music of the Peoples of the World, by William Alves (Schirmer, 2005)

The Rough Guide to World Music, Vols. 1–3, by Broughton and Ellington (Rough Guides, 2006)

World Music Press
www.worldmusicpress.com

World Musics in Context: A Comprehensive Survey of the World's Major Musical Cultures,

by Peter Fletcher (Oxford University Press, 2004)

Worlds of Music: An Introduction to the Music of the World's Peoples, by Jeff Todd Titon (Schirmer, 2004)

General Info

World Music Central
www.worldmusiccentral.org

World Music Network
www.worldmusic.net

Concerts, Festivals, and Expositions

San Francisco World Music Festival
www.sfworldmusicfestival.org

WOMEX
www.womex.de

World Music
www.worldmusic.org

World Music Institute
www.heartheworld.org

Education

The Archive of World Music
http://hcl.harvard.edu/libraries/loebmusic/collections/archive.html

The Society for Ethnomusicology
http://webdb.iu.edu/sem/scripts/home.cfm

Smithsonian Folkways
www.folkways.si.edu/index.html

Traditional World Music Recordings
www.medieval.org/music/world.html

UCLA Ethnomusicology Archive
www.ethnomusic.ucla.edu/archive/preservation.htm

Music and Dance Classes

New York, NY

Capoeira Angola Center
www.panix.com/~tishotto/capoeira

Chinese Folk Dance Company
www.chinesedance.org

Djoniba Dance & Drum Centre, Inc.
(African drum and dance)
www.djoniba.com/africandance.html

Fazil Studio
(Merengue, samba, tango)
www.fazilstudio.com

Harkness Dance Center
(Afro-Caribbean and others)
www.92y.org

Miami, FL

Bozenka's Bellydance Academy
(Middle Eastern dances including belly dance)
www.emerald-dreams.com

Gypsy Fire Dance
(Flamenco and belly dance)
www.gypsyfire.org

World Music Sources continued

Music and Dance Classes

Washington, DC

Sahara Dance
(Middle Eastern dance)
http://saharadance.com

Cleveland, OH

The Dance Studio
(African and salsa)
http://moodiedance.com/

Chicago, IL

Arabesque Chicago
(Middle Eastern, belly dance)
www.arabesquechicago.com

Chicago Dance
(Latin, salsa, mambo)
www.chicagodance.com

Old Town School of Folk Music
(African, Latin, Middle Eastern, Caribbean)
www.oldtownschool.org

Minneapolis, MN

Four Seasons Dance Studio (Latin)
www.fourseasonsdance.com

Houston, TX

Sonia's World of Dance
(Salsa, *merengue*, polka)
www.houston-dance-lessons.com

Denver, CO

Belly Dance with Sadie
www.bedouinprincess.com

Bollywood West
(Indian)
www.bollywoodwest.com/news_media.htm

Seattle, WA

Gansango Dance Company
(African)
www.gansango.com

Seattle Belly Dancing Class
www.visionarydance.com

Portland, OR

Viscount Dance Studios
(Tango, samba, salsa, Caribbean)
www.viscountstudios.com

San Francisco, CA

Salsa Crazy
www.salsacrazy.com

Los Angeles, CA

Dance Arts Academy
(West African, Irish, salsa, tango)
www.danceartsacademy.com

Josie Neglia Dance Academy
(*Reggaeton*, salsa, cha-cha, Latin)
www.jnda.us

Notes

Notes

400,000
STUDENTS
READ
MUSIC ALIVE!™
Music Education for Today's Generation
DO YOURS?
Grab their attention with the hits they love while delivering a standards-based curriculum with a subscription to *Music Alive!*
www.musicalive.com